Mechanics and Geometry of Enriched Continua

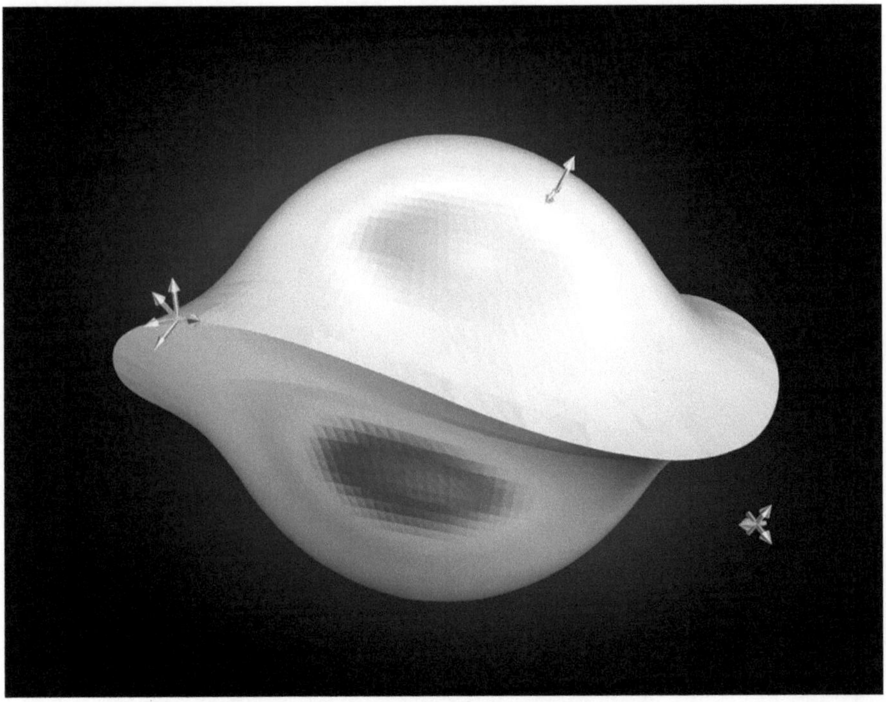

Luis Espath

Mechanics and Geometry of Enriched Continua

Springer

Luis Espath
School of Mathematical Sciences
University of Nottingham
Nottingham, UK

ISBN 978-3-031-28933-0 ISBN 978-3-031-28934-7 (eBook)
https://doi.org/10.1007/978-3-031-28934-7

This Springer imprint is published by the registered company Springer Nature Switzerland AG
The registered company address is: Gewerbestrasse 11, 6330 Cham, Switzerland

Preface

This study resulted from my Habilitation thesis, *Mechanics and Geometry of Enriched Continua*, defended in the Department of Mathematics at RWTH Aachen University, Germany, in addition to the research paper [1] published shortly after. The prerequisite knowledge, tensor analysis, kinematics, and mechanical principles including thermodynamics, to follow this study may be found in the book by Gurtin et al. [2]. However, other books on continuum mechanics may suffice, such as those by Malvern [3] or Gurtin [4]. I assume the material herein presented would be of interest to students enrolled in programs of Applied Mathematics and Aerospace, Civil, and Mechanical Engineering.

This study is not a comprehensive treatment of continuum mechanics. This represents my standpoint when it comes to enriched continua, specifically gradient theories, where I completely disregard incompatible theories. The last chapter, though, presents a special theory to add an additional flavor to the enrichment of the continuum medium, a theory that does not fall into the category of gradient or incompatible theories.

Based upon the pillars of continuum mechanics, the *Mechanics and Geometry of Enriched Continua* presents a standalone treatment of gradient theories for motion and an arbitrary number of transition layers where special attention is paid to incompressible fluid flows. In general, these transition layers may represent components and/or phases. This study is split into five parts, in addition to the conclusions. Part I motivates the study and introduces the underlying mathematical tools needed to follow the mathematical content. Part II develops the principle of virtual powers to arrive at the field equations, frame-indifference, action-reaction principle, surface balance principles, and the virtual power principle on surfaces. Part III covers the classical thermodynamics as well as some special topics on inner products and the representation theorem. Part IV discusses environmental balances and imbalances to derive thermodynamically consistent boundary conditions. Part V introduces a type of continuum theory where the boundary conditions are defined by additional field equations on surfaces, which end up being coupled with the underlying field equations in the bulk. Lastly, in Part VI, I discuss the conclusions and limitations, where I also present a brief synopsis of this study.

You can reach out to me[1] if you find any typographical errors, have suggestions, criticism, or just would like to discuss the content of this study.

Nottingham, UK Luis Espath

[1] luis.espath@nottingham.ac.uk.

Acknowledgments

I am indebted to many colleagues with whom I had exhaustive discussions in which all of them gave me valuable ideas, constructive comments, and encouragement. Also thanks to my son Luca and Gabriel Nogueira de Castro, who taught me Blender to generate the figures for this study.

Special thanks, in alphabetical order, to Alexander Litvinenko, Anne Boschman, Arved Bartuska, Athanasios Tzavaras, Bonbien Varga, Chiheb Ben Hammouda, Eliot Fried, Håkon Hoel, Harald van Brummelen, Michael Westdickenberg, Nadhir Ben Rached, Nicolas Labanda, Raúl Tempone, Santiago Clavijo, Sebastian Krumscheid, Sophia Wiechert, Victor Calo, and my kids and wife.

Contents

Part I
Point of Departure and Machinery

Chapter 1
Introduction

A historical note

Let me embark on briefly revisiting some pieces of the work by Leonardo da Vinci (1452–1519). Inspired by the chaotic nature of fluid flows and more than a century before the Newtonian formalism (1672–1727),[1] Leonardo da Vinci studied fluid flows, see Fig. 1 and also the work by Kemp [5] on 'Leonardo da Vinci's laboratory: studies in flow'. Although Archimedes' work (287–212 BC) on the laws of buoyancy may be the first known study of fluid statics, as acknowledge by Kemp, it was Leonardo's sense of *turbolenza*[2] and kinematics the first milestone in fluid dynamics. Probably, his most celebrated work on fluid flows is that of the flow and vortex formation in the aorta. He identified the vortices in the sinus of Valsalva as the mechanism responsible for the closing of the valve, as acknowledged by Marusic & Broomhall [6].

Enriched continua

Enriched continuum is a broad concept that traces back to the Cosserat brothers [7] in 1909. The *Cosserat medium*[3] is endowed with translational and rotational degrees of freedom to describe its kinematics. However, it was Voigt [9] in 1887 who manifested the need of a *surface-couple traction* between adjacent parts, aside from the classical *surface tractions* arising on Euler–Cauchy cuts, as cited by Ganghoffer [10].

---

[1] Quoting Gurtin et al. [2]: "Those who believe the notion of force is obvious should read the scientific literature of the period following Newton. Truesdell (1966) notes that *D'Alembert spoke of Newtonian forces as obscure and metaphysical beings, capable of nothing but spreading darkness over a science clear by itself*, while Jammer (1957, pp. 209, 215) paraphrases a remark of Maupertis, *we speak of forces only to conceal our ignorance*, and one of Carnot, *an obscure metaphysical notion, that of force*."

[2] *Turbolenza*: turbulence in Italian.

[3] For an overview of enriched continua theories, the reader is referred to the book edited by Maugin & Metrikine [8], *one hundred years after the Cosserats*.

L. Espath, *Mechanics and Geometry of Enriched Continua*,
https://doi.org/10.1007/978-3-031-28934-7_1

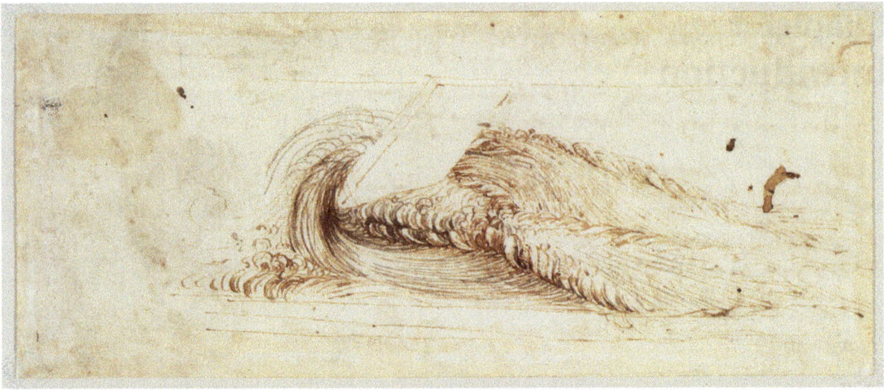

A. Flow around an obstacle

B. Old man seated and flow around obstacles with notes

Fig. 1 **A** Leonardo da Vinci, Verso: Studies of flowing water c.1510-13, Royal Collection Trust/©His Majesty King Charles 2023 **B** Leonardo da Vinci, Recto: Studies of water, and a seated old man c.1512-13, Royal Collection Trust/©His Majesty King Charles 2023

enlargethispage-6ptFurthermore, according to Edelen [11], Voigt also thought and elaborated the idea of nonlocalities to augment the hypothesis of Euler–Cauchy. Conversely, Duhem [12] in 1893 expressed the need for additional degrees of freedom, *directors*, to describe internal rotations, as acknowledged by Ericksen & Truesdell [13]. Duhem postulated that the state of the matter at a point is also characterized by a direction. In addition to the work by the Cosserat brothers, Ericksen [14] also exploited directors to describe internal rotations in liquid crystals. Finally, to extend nonclassical theories, Toupin [15, 16] introduced the notion and formalism of *gradient theories*. Gradient theories in continuum mechanics account for long-distance interactions that are necessary to model materials that exhibit multiple dominant length scales. Toupin [15, 16] introduced and studied second-grade materials with couple effects to the elasticity theory. Later, Mindlin & Eshel [17] and Germain [18, 19] extended these ideas by using the virtual work machinery. Analogously, using the principle of virtual power, Fried & Gurtin [20] generalized Toupin's work to account for general second-grade materials, including fluid flows. Noll & Virga [21] analyzed edge interactions, studying the concept of forces distributed over edges, as initially described by Toupin [15]. Dell'Isola & Seppecher [22] formalized and studied the concept of edge tractions. In 1989, Fosdick [23], proved the existence of a linear transformation on surfaces that transform the unit normal into the surface traction, namely, the Cauchy stress. He used a variational approach. In 2016, using similar arguments, Fosdick [24] also proved the existence of a linear transformation on edges that transforms the unit normal and unit tangent-normal into the edge traction. Building from here, Fosdick systematically generalized continuum theories. Oversimplifying the literature, one is led to find that enriched continua bridge different length scales. That is, these theories upscale the influence of microstructure to the underlying macroscopic behavior of materials. Thus, micromorphic and gradient continuum theories are a homogenized type of continua that upscale information from the underlying microstructure. However, in the literature of the period by Woldemar Voigt, Pierre Duhem, François Nicolas, and Eugène-Maurice-Pierre Cosserat, one is led to believe that this definition of enriched continua is not what they meant. Likewise, in the theory of liquid crystals with variable degree of orientation by Ericksen, the variable degree of orientation is intimately linked to the motion and description of defects.

In panels (A), (B), (C), and (D) of Fig. 2, the pioneers of nonlocalities in continuum theories, Woldemar Voigt, Pierre Duhem, François Nicolas, and Eugène-Maurice-Pierre Cosserat are portrayed, respectively.

Transition layers

Van der Waals [25, *translation*] may have made the first reference to diffuse transition layers in fluids. He proposed a continuous equation of state describing the thermodynamic properties of water in both liquid and vapor phases. His fundamental hypothesis for this fluid model is that the interface has a non-vanishing thickness. Moreover, across the thickness, fluid properties are smoothly distributed. These smooth changes in properties are usually addressed by an order parameter, a parameter that describes the degree of order across the boundaries in a phase transition system. Korteweg

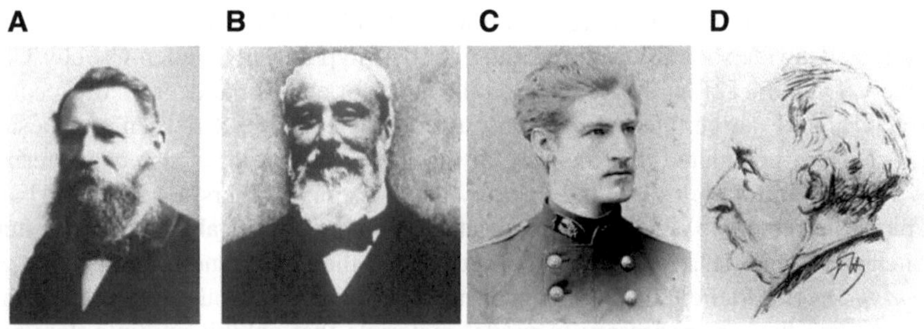

Fig. 2 Historical figures related to enriched continuum theories: **A** Photograph of Woldemar Voigt (Göttingen University History—Portraits, Lower Saxony State and University Library Göttingen, approx. 1890, licensed under CC-BY 4.0) **B** Photograph of Pierre Duhem (Wikimedia Commons, before 1916, licensed under CC-PD-mark 1.0) **C** Photograph of François Nicolas Cosserat (Wikimedia Commons, before 1914, licensed under CC-PD-mark 1.0) **D** Photograph of Eugène-Maurice-Pierre Cosserat (Wikimedia Commons, 1929, licensed under CC-PD-mark 1.0)

[26] derived a capillarity stress tensor arising in Van der Waals fluid. Dunn & Serrin [27] augmented the mechanical version of the second law of thermodynamics with the 'interstitial working' to describe the Korteweg-type of transition in a thermodynamically consistent manner. These types of continuum theories are also termed as diffuse-interface theories as described by Anderson et al. [28]. In a collection of papers, Fried & Gurtin [29, 30] and Gurtin [31] describe the evolution of transition layers from a mechanistic standpoint. In [29] the notion of balance laws for accretive forces is introduced along with the accretive stress, a vectorial quantity, and the accretive internal and external forces, scalar quantities. The 'accretive'[4] terminology was shortly thereafter abandoned in [30], and in [31] the 'microforces' and 'microstresses' terminology was adopted. However, the underlying physical meaning is the same. Fried & Gurtin derived the 'generalized Allen–Cahn–Ginzburg–Landau' and the 'generalized Cahn–Hilliard' equations by introducing the balance of microforces. Their monumental work makes explicit the underlying 'forces' that dictate the evolution of phase fields which describe transition layers. In the same manner that forces and stresses are related to motion and deformation, microforces and microstresses are associated with the phase-field evolution (of one phase with respect to others) and growth (of one phase at the expense of others). Aligned with the works by Fried & Gurtin, Espath et al. [32] extended and generalized these ideas to second-grade phase-field theories by capitalizing on the works by Fosdick & Virga [23] and Fosdick [24]. Also, Espath et al. [33] used the principle of virtual powers for second-grade phase-field theories to derive the 'generalized Swift–Hohenberg' equation and the 'generalized phase-field crystal' equation. Lastly, when it comes to capillarity effects, another relevant work is that by Gurtin et al. [34]. While Dunn and

[4] Here, I quote Fried & Gurtin [29]: "*Since it is the accretion of material of one phase at the expense of another that characterizes phase transitions, we use the adjective 'accretive' to describe this system of forces as well as its associated balance law.*"

A B

Fig. 3 Historical figures related to the criticality phenomenon: **A** Photograph of Johannes Diderik van der Waals (Wikimedia Commons, before 1923, licensed under CC-PD-mark 1.0) **B** Photograph of Diederik Johannes Korteweg (Archief Amsterdam 010097008741. By Molkenboer, Theodorus (1796–1863), Tresling en Co., Collectie Stadsarchief Amsterdam: tekeningen en prenten, 1897, Copyright free)

Serrin enforce thermodynamic consistency by accounting for an 'interstitial working' in the energy equation, Gurtin et al. capitalized on the microforces' framework to achieve an analogous result.

In panels (A) and (B) of Fig. 3, the pioneers of criticality in continuum theories, Johannes Diderik van der Waals and Diederik Johannes Korteweg are portrayed, respectively.

This study

I propose a continuum theory for enriched continua. Restricting attention to gradient theories, I study motion exhibiting an arbitrary number of enriched transition layers. As a point of departure, I account for multiple continuum kinematic processes, including temporal changes in motion and temporal changes in various phase fields, which describe the transition layers. The velocity is denoted by $\boldsymbol{v}$ while the phase fields by $\{\varphi^\alpha\}_{\alpha=1}^n$. I base my treatment on the principle of virtual powers[5] to determine the explicit form of the fields present in the external and internal virtual

[5] Other alternatives to the principle of virtual powers have been explored to achieve similar results. Some of these are the *Euclidean frame indifference* of the energy equation by Green & Rivlin [35, 36] and Svendsen & Bertram [37, 38]; and likewise, the *rate-variational* approach was proposed by Svendsen [39, 40] and Svendsen et al. [41]. These continuum frameworks are usually coupled with dissipation principles, and the continuum is endowed with microstructure, namely, transition

powers and obtain the pointwise version of the balances of microforces and forces. The actual kinematic processes are related to temporal changes in the thermodynamic state variables, such as temperature, motion, and phase fields. In the internal virtual power, I incorporate the first, second, and third gradients of phase fields and velocity. These gradients arise by considering various kinematic processes which are intimately connected to the level of the arbitrariness of the Euler–Cauchy cuts. The surface defining the Euler–Cauchy cut may lose its smoothness along a curve, namely, a junction-edge. A junction-edge may also lose its smoothness at a point, namely, a junction-point. On each geometrical feature, interactions between adjacent parts of the body are developed. These interactions are described by microtractions, tractions, microforces, and forces, which are power conjugate to independent kinematic processes. The pointwise balances are then integrated on an arbitrary part to arrive at the partwise balances of microforces, forces, microtorques, and torques. Next, I postulate surface balances of microforces, forces, microtorques, and torques. Additionally, I postulate the principle of virtual power on surfaces. Then, the first and second laws of thermodynamics with the power balance provide suitable and consistent choices for the constitutive equations. Finally, the complementary balances, namely, the balances on surfaces, are tailored to coincide with different parts of the boundaries of the body. These surface balances are then called environmental surface balances and aid in determining suitable and consistent boundary conditions. Ultimately, the environmental surface power balance is relaxed to yield an environmental surface imbalance of powers, rendering a more general type of boundary condition. This continuum framework is based on, but also extends and complements, the work by Fried & Gurtin [29, 30], Gurtin [31, 42], Podio-Guidugli [43], Carillo et al. [44], Fosdick & Virga [23], Fosdick [24], Espath et al. [33], Espath & Calo [32], Espath [45], and Clavijo et al. [46, 47].

As notational agreement, with different font types I denote scalars a, vectors $\boldsymbol{a}$, second- $\mathbf{A}$, third- $\mathbb{A}$, and fourth-order $\boldsymbol{\mathbb{A}}$ tensor fields. $(\cdot)^{\top}$ is the minor right transposition, that is, transposition on the last two indices regardless of the order of the tensor, $(\cdot)^{\perp}$ is the minor left transposition, that is, transposition on the first two indices regardless of the order of the tensor, while $(\cdot)^{\dashv}$ is the transposition between the last index and all the rest and $(\cdot)^{\vdash}$ is the transposition between the first index and all the rest. Lastly, $\cdot[\cdot]$ indicates a complete contraction of tensors of a different order, for instance, $\mathbf{A}[\boldsymbol{b}]$ is a vector, $\mathbb{A}[\boldsymbol{b}]$ a second-order tensor, $\mathbb{A}[\mathbf{B}]$ a vector, and $\boldsymbol{\mathbb{A}}[\mathbf{B}]$ is a second-order tensor.

Here, $\mathcal{B}$ denotes a region of a three-dimensional point space $\mathcal{E}$ where $\mathcal{P} \subseteq \mathcal{B}$ is an arbitrary subregion of $\mathcal{B}$ with a closed surface boundary $\partial\mathcal{P}$ oriented by an outward unit normal $\boldsymbol{n}$ at $\boldsymbol{x} \in \partial\mathcal{P}$. The surface $\partial\mathcal{P}$ may lose its smoothness along a curve, namely, junction-edge $\mathcal{C}$. In analyzing a neighborhood of a junction-edge $\mathcal{C}$, two smooth surfaces $\partial\mathcal{P}^{\pm}$ are defined. Thus, the limiting unit normals of $\partial\mathcal{P}^{\pm}$ at $\mathcal{C}$ are denoted by the tuple $\{\boldsymbol{n}^{+}, \boldsymbol{n}^{-}\}$, which characterizes the junction-edge $\mathcal{C}$.

---

layers. In the rate-variational approach, the actual internal and external powers are expressed in terms of rate-potential forms. To date, these approaches have been employed for gradient theories endowed with up to second gradients in the motion and first gradients in the phase field.

Equivalently, the limiting outward unit tangent-normal of $\partial\mathcal{P}^{\pm}$ at $\mathcal{C}$ are $\{\boldsymbol{\nu}^{+}, \boldsymbol{\nu}^{-}\}$. Additionally, $\mathcal{C}$ is oriented by the unit tangent $\boldsymbol{\sigma} := \boldsymbol{\sigma}^{+}$ such that $\boldsymbol{\nu}^{+} := \boldsymbol{\sigma}^{+} \times \boldsymbol{n}^{+}$. A junction-edge may also lose its smoothness at a point, namely, junction-point $\mathcal{O}$. In analyzing a neighborhood of a junction-point $\mathcal{O}$, two smooth junction-edges $\mathcal{C}^{\pm}$ are defined. The limiting unit tangents of $\mathcal{C}^{\pm}$ at $\mathcal{O}$ are denoted by the tuple $\{{}^{+}\boldsymbol{\sigma}, {}^{-}\boldsymbol{\sigma}\}$, which characterizes the junction-point $\mathcal{O}$. Furthermore, the body $\mathcal{B}$ and all its parts are open sets in $\mathcal{E}$.

We reserve the notion known as partwise balances for balance integrals on volumes, as customary in continuum mechanics. The notion of pointwise balances makes reference to the localization of partwise balances which renders the partial differential equations. Lastly, the field equations are the pointwise balances of motion and evolution of the phase fields. In a similar fashion, we also have the notion of partwise surface balances for balance integrals on surfaces while pointwise surface balances arise by localization.

1 Fields, Hyperfields & Couple-Fields

For the sake of presentation, although I use the terminology *field* with the usual connotation found in calculus books, I distinguish it into fields, hyperfields, and couple-fields. To unfold the implications of considering arbitrary parts $\mathcal{P}$ that lack smoothness at a junction-edge $\mathcal{C}$ and at a junction-point $\mathcal{O}$ arising in this theory, I start discussing the fields, hyperfields, and couple-fields developed on different geometrical entities of an arbitrary part $\mathcal{P}$ such as surfaces $\mathcal{S}$, curves $\mathcal{C}$, and points $\mathcal{O}$. I first recall that, originally, on Euler–Cauchy cuts (a surface $\mathcal{S} \subset \mathcal{B}$) the surface traction is assumed to depend upon one geometrical descriptor, namely, the unit normal $\boldsymbol{n}$. Aside from the unit normal, Toupin [15, 16] augmented this hypothesis and included yet another geometrical descriptor, namely, the curvature tensor $\mathbf{K} := -\mathrm{grad}_{\mathcal{S}}\boldsymbol{n}$, the negative surface gradient of the unit normal, along with the notion of edge tractions. Noll & Virga [21], Dell'Isola & Seppecher [22], Fried & Gurtin [20], and Fosdick [24] capitalized on Toupin's work to depict the interplay between the edge traction and the surface traction. Also, it was shown that the edge traction depends upon the unit normal $\boldsymbol{n}$ and unit tangent-normal $\boldsymbol{\nu}$. Here, I consider yet another geometrical descriptor to describe microtractions and tractions aside from the tuple $\{\boldsymbol{n}, \boldsymbol{\nu}, \mathbf{K}\}$, namely, the second curvature tensor $\mathbb{K} := -\mathrm{grad}_{\mathcal{S}}^{2}\boldsymbol{n}$, the negative second surface gradient of the unit normal. This additional geometrical descriptor already appeared in the works by Podio-Guidugli [43] and Carillo et al. [44] in the study of hyperelastic materials, also in the realm of enriched continua. Next, let me describe these fields, hyperfields, and couple-fields.

Figure 4 depicts the level of arbitrariness of a part considered in this study. The unit normal $\boldsymbol{n}$, unit tangent-normal $\boldsymbol{\nu}$, and unit tangent $\boldsymbol{\sigma}$ fields may be discontinuous. In this figure, the surface $\mathcal{S}$ lacks smoothness at a junction-edge $\mathcal{C}$. At $\mathcal{C}$, the unit normal $\boldsymbol{n}$ and unit tangent-normal $\boldsymbol{\nu}$ fields are discontinuous. The junction-edge $\mathcal{C}$ lacks smoothness at a junction-point $\mathcal{O}$. At $\mathcal{O}$, the unit tangent $\boldsymbol{\sigma}$ field is discontinuous.

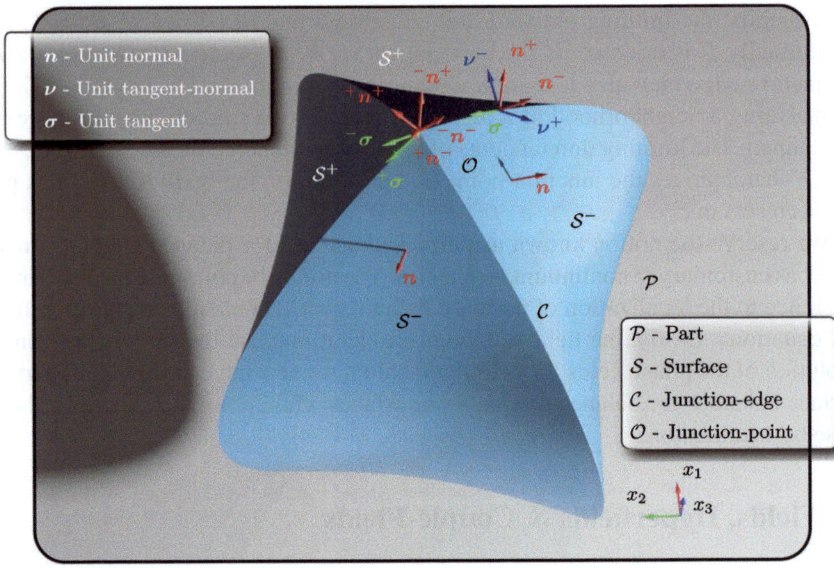

Fig. 4 An arbitrary part $\mathcal{P}$. The surface $\mathcal{S}$ lacks smoothness at a junction-edge $\mathcal{C}$. At $\mathcal{C}$, the unit normal $\boldsymbol{n}$ and unit tangent-normal $\boldsymbol{\nu}$ fields are discontinuous. The junction-edge $\mathcal{C}$ lacks smoothness at a junction-point $\mathcal{O}$. At $\mathcal{O}$, the unit tangent $\boldsymbol{\sigma}$ field is discontinuous

We are now in a position to define the internal interactions, provoked by the αth transition layer and motion, in addition to external agents acting on $\mathcal{B}$ as follows.

(i) *The αth external microforce $\gamma^\alpha := \gamma^\alpha(\boldsymbol{x}, t)$ and external force $\boldsymbol{b} := \boldsymbol{b}(\boldsymbol{x}, t)$* represent, respectively, a body microforce and a body force per unit mass, acting on the body $\mathcal{B}$. Letting $\varrho := \varrho(\boldsymbol{x}, t)$ be the density, $\boldsymbol{b}$ represents both the net inertial and noninertial body forces. Thus, $\boldsymbol{b} := \boldsymbol{b}^{\mathrm{ni}} + \boldsymbol{b}^{\mathrm{in}}$, where $\boldsymbol{b}^{\mathrm{ni}}$ is the noninertial body force and $\boldsymbol{b}^{\mathrm{in}}$ the inertial body force. I let $\boldsymbol{b}^{\mathrm{in}} := -\varrho\dot{\boldsymbol{v}}$ be a function of the density ϱ and the acceleration $\dot{\boldsymbol{v}}$. The body microforces and body force fields are developed by external causes, outside of $\mathcal{B}$.

Common to all continuum theories, the notion of stress accompanies the notion of relative motion between distinct particles of $\mathcal{B}$. Conversely, the notion of microstress arises as an act in response to changes in the internal microstructure, which is described by the phase fields. Guided by the presence of the first, second, and third gradients of the velocity and the phase fields, I here consider stresses-, hyperstresses-, and supra hyperstresses-like fields as follows.

(ii) *The αth microstress $\boldsymbol{\xi}^\alpha := \boldsymbol{\xi}^\alpha(\boldsymbol{x}, t)$ and stress $\mathbf{T} := \mathbf{T}(\boldsymbol{x}, t)$ fields* represent, respectively, a microforce per unit length and a force per unit area (or an energy per unit volume/length and an energy per unit volume).

(iii) *The αth hypermicrostress $\Xi^\alpha := \Xi^\alpha(\boldsymbol{x}, t)$ and hyperstress $\mathbb{T} := \mathbb{T}(\boldsymbol{x}, t)$ fields* represent, respectively, a microforce per unit length/length and a force per unit area/length (or an energy per unit volume/area and an energy per unit volume/length).

(iv) *The αth supra hypermicrostress* $\mathbb{X}^\alpha := \mathbb{X}^\alpha(\boldsymbol{x}, t)$ *and second hyperstress* $\mathbf{T} := \mathbb{T}(\boldsymbol{x}, t)$ *fields* represent, respectively, a microforce per unit length/area and a force per unit area/area (or an energy per unit volume/volume and an energy per unit volume/area).

While $\boldsymbol{\xi}^\alpha$, $\boldsymbol{\Xi}^\alpha$, and $\mathbb{X}^\alpha$ characterize forces transmitted across surfaces, $\mathbf{T}$, $\mathbb{T}$, and $\mathbf{T}$ characterize forces transmitted on surfaces. Here, the fields $\boldsymbol{\xi}^\alpha$ and $\mathbf{T}$ constitute the basic ingredients of a classical theory while the hyperfields $\{\boldsymbol{\Xi}^\alpha, \mathbb{T}\}$ and $\{\mathbb{X}^\alpha, \mathbf{T}\}$ constitute additional ingredients of second- and third-grade continuum theories, respectively.

The microtractions, tractions, concentrated microforces, and concentrated forces developed on Euler–Cauchy cuts arising in this continuum theory are as follows.

(i) *The αth surface microtraction* $\xi^\alpha_{\mathcal{S}} := \xi^\alpha_{\mathcal{S}}(\boldsymbol{x}, t; \boldsymbol{n}, \mathbf{K}, \mathbb{K})$ *and the surface traction* $\boldsymbol{t}_{\mathcal{S}} := \boldsymbol{t}_{\mathcal{S}}(\boldsymbol{x}, t; \boldsymbol{n}, \mathbf{K}, \mathbb{K})$ represent, respectively, a microforce per unit length and a force per unit area, acting on an oriented surface $\mathcal{S} \subset \mathcal{B}$ at $\boldsymbol{x} \in \mathcal{S}$. Both depend on $\mathcal{S}$ through the tuple $\{\boldsymbol{n}, \mathbf{K}, \mathbb{K}\}$. The αth surface microtraction and surface traction expend power per unit area, respectively, conjugate to $\dot{\varphi}^\alpha$ and $\boldsymbol{v}$ on $\mathcal{S}$, and are developed by the contact of a surface $\mathcal{S}$ with the adjacent parts of $\mathcal{B}$.

(ii) *The αth junction-edge microtraction* $\xi^\alpha_{\mathcal{C}} := \xi^\alpha_{\mathcal{C}}(\boldsymbol{x}, t; \boldsymbol{n}^\pm, \mathbf{K}^\pm)$ *and the junction-edge traction* $\boldsymbol{t}_{\mathcal{C}} := \boldsymbol{t}_{\mathcal{C}}(\boldsymbol{x}, t; \boldsymbol{n}^\pm, \mathbf{K}^\pm)$ represent, respectively, a microforce per unit length/length and a force per unit area/length, acting on a junction-edge $\mathcal{C}$ of a nonsmooth oriented surface $\mathcal{S} \subset \mathcal{B}$ at $\boldsymbol{x} \in \mathcal{C} \subset \mathcal{S}$. Both depend on $\mathcal{C}$ and $\mathcal{S}$ through the tuple $\{\boldsymbol{n}^\pm, \mathbf{K}^\pm\}$ defined as their limiting values of each smooth part of $\mathcal{S}$ at $\mathcal{C}$. The αth junction-edge microtraction and the junction-edge traction expend power per unit length, respectively, conjugate to $\dot{\varphi}^\alpha$ and $\boldsymbol{v}$ on $\mathcal{C}$, and are developed by the contact of a junction-edge $\mathcal{C}$ with the adjacent parts of $\mathcal{B} \setminus \mathcal{S}$.

(iii) *The αth junction-point microforce* $\xi^\alpha_{\mathcal{O}} := \xi^\alpha_{\mathcal{O}}(\boldsymbol{x}, t; {}^\pm\boldsymbol{n}^\pm)$ *and the junction-point force* $\boldsymbol{t}_{\mathcal{O}} := \boldsymbol{t}_{\mathcal{O}}(\boldsymbol{x}, t; {}^\pm\boldsymbol{n}^\pm)$ represent, respectively, a concentrated microforce per unit length/area and a concentrated force per unit area/area, acting at a junction-point $\mathcal{O}$ of a nonsmooth oriented junction-edge $\mathcal{C} \subset \mathcal{S}$. Both depend on $\mathcal{C}$ and $\mathcal{S}$ through the tuple $\{{}^\pm\boldsymbol{n}^\pm\}$. The αth junction-point microforce and the junction-point force expend power, respectively, conjugate to $\dot{\varphi}^\alpha$ and $\boldsymbol{v}$ at $\mathcal{O}$, and are developed by the contact of a junction-point $\mathcal{O}$ with the adjacent parts of $\mathcal{B} \setminus \{\mathcal{S} \cup \mathcal{C}\}$.

Owing to the nature of this theory in which I consider the existence of hyperfields—namely, hypermicrostresses, a hyperstress, supra hypermicrostresses, and a supra hyperstress—hypermicrotractions and hypertractions appearing in the principle of virtual power and are described in what follows.

(i) *The αth surface hypermicrotraction* ${}^2\xi^\alpha_{\mathcal{S}} := {}^2\xi^\alpha_{\mathcal{S}}(\boldsymbol{x}, t; \boldsymbol{n}, \mathbf{K})$ *and the surface hypertraction* ${}^2\boldsymbol{t}_{\mathcal{S}} := {}^2\boldsymbol{t}_{\mathcal{S}}(\boldsymbol{x}, t; \boldsymbol{n}, \mathbf{K})$ represent, respectively, a microforce per unit length/length and a force per unit area/length, acting on an oriented surface $\mathcal{S} \subset \mathcal{B}$ at $\boldsymbol{x} \in \mathcal{S}$. Both depend on $\mathcal{S}$ through the tuple $\{\boldsymbol{n}, \mathbf{K}\}$. The αth surface hypermicrotraction and surface hypertraction expend power, respectively, conjugate to $\partial_n\dot{\varphi}^\alpha$ and $\partial_n \boldsymbol{v}$ on $\mathcal{J}$ by the contact of adjacent parts of $\mathcal{B}$.[6]

(ii) *The αth surface supra-hypermicrotraction* ${}^3\xi^\alpha_{\mathcal{S}} := {}^3\xi^\alpha_{\mathcal{S}}(\boldsymbol{x}, t; \boldsymbol{n})$ *and the surface supra-hypertraction* ${}^3\boldsymbol{t}_{\mathcal{S}} := {}^3\boldsymbol{t}_{\mathcal{S}}(\boldsymbol{x}, t; \boldsymbol{n})$ represent, respectively, a microforce per unit length/area and a force per unit area/area, acting on an oriented surface $\mathcal{S} \subset \mathcal{B}$ at $\boldsymbol{x} \in \mathcal{S}$. Both depend on $\mathcal{S}$ through $\boldsymbol{n}$. The αth surface supra-hypermicrotraction and surface supra-hypertraction expend power, respectively, conjugate to $\partial^2_n\dot{\varphi}^\alpha$ and $\partial^2_n \boldsymbol{v}$ on $\mathcal{S}$ by the contact of adjacent parts of $\mathcal{B}$.

(iii) *Decomposed into two components, the αth junction-edge hypermicrotraction* $\{{}^2_n\xi^\alpha_{\mathcal{C}} := {}^2_n\xi^\alpha_{\mathcal{C}}(\boldsymbol{x}, t; \boldsymbol{n}^\pm),\ {}^2_\nu\xi^\alpha_{\mathcal{C}} := {}^2_\nu\xi^\alpha_{\mathcal{C}}(\boldsymbol{x}, t; \boldsymbol{n}^\pm)\}$ *and the junction-edge hypertraction* $\{{}^2_n\boldsymbol{t}_{\mathcal{C}} := {}^2_n\boldsymbol{t}_{\mathcal{C}}(\boldsymbol{x}, t; \boldsymbol{n}^\pm),\ {}^2_\nu\boldsymbol{t}_{\mathcal{C}} := {}^2_\nu\boldsymbol{t}_{\mathcal{C}}(\boldsymbol{x}, t; \boldsymbol{n}^\pm)\}$ represent, respectively, a microforce per unit length/area and a force per unit area/area, acting on a junction-edge $\mathcal{C}$ of a nonsmooth oriented surface $\mathcal{S} \subset \mathcal{B}$ at $\boldsymbol{x} \in \mathcal{C} \subset \mathcal{S}$. Both depend on $\mathcal{C}$ and $\mathcal{S}$ through the tuple $\{\boldsymbol{n}^\pm\}$. The αth junction-edge hypermicrotraction and junction-edge hypertraction expend power, respectively, conjugate to $\{\partial_n\dot{\varphi}^\alpha, \partial_\nu\dot{\varphi}^\alpha\}$ and $\{\partial_n \boldsymbol{v}, \partial_\nu \boldsymbol{v}\}$ on $\mathcal{C}$ by the contact of adjacent parts of $\mathcal{B} \setminus \mathcal{S}$.

Another important set of microtractions and tractions are the couple-fields—namely, couple-microtractions and couple-tractions—appearing in the balances of microtorques and torques and are described in what follows.

(i) *The αth surface-couple microtraction* $\varpi^\alpha_{\mathcal{S}} := \varpi^\alpha_{\mathcal{S}}(\boldsymbol{x}, t; \boldsymbol{n}, \mathbf{K})$ *and the surface-couple traction* $\boldsymbol{m}_{\mathcal{S}} := \boldsymbol{m}_{\mathcal{S}}(\boldsymbol{x}, t; \boldsymbol{n}, \mathbf{K})$ represent, respectively, a microtorque per unit length and a torque per unit area, acting on an oriented surface $\mathcal{S} \subset \mathcal{B}$ at $\boldsymbol{x} \in \mathcal{S}$. Both depend on $\mathcal{S}$ through the tuple $\{\boldsymbol{n}, \mathbf{K}\}$. The αth surface-couple microtraction and surface-couple traction are also developed by the contact of a surface $\mathcal{S}$ with the adjacent parts of $\mathcal{B}$.

(ii) *The αth junction-edge-couple microtraction* $\varpi^\alpha_{\mathcal{C}} := \varpi^\alpha_{\mathcal{C}}(\boldsymbol{x}, t; \boldsymbol{n}^\pm)$ *and the junction-edge-couple traction* $\boldsymbol{m}_{\mathcal{C}} := \boldsymbol{m}_{\mathcal{C}}(\boldsymbol{x}, t; \boldsymbol{n}^\pm)$ represent, respectively, a microtorque per unit length/length and a torque per unit area/length, acting on a junction-edge $\mathcal{C}$ of a nonsmooth oriented surface $\mathcal{S} \subset \mathcal{B}$ at $\boldsymbol{x} \in \mathcal{C} \subset \mathcal{S}$. Both depend on $\mathcal{C}$ and $\mathcal{S}$ through the tuple $\{\boldsymbol{n}^\pm\}$. The αth junction-edge-couple microtraction and the junction-edge couple-traction are developed by the contact of a junction-edge $\mathcal{C}$ with the adjacent parts of $\mathcal{B} \setminus \mathcal{S}$.

Lastly, while microtractions and couple-microtractions characterize, respectively, forces and torques transmitted across surfaces, tractions, and couple-tractions characterize, respectively, forces and torques transmitted on surfaces.

2 The Principle of Virtual Powers' Formalism

For the point I wish to make, consider the continuum theory for 'simple materials' with a single phase and velocity fields describing the kinematics. In this scenario,

[6] The symbol ∂_y is the conventional partial derivative with respect to a generic variable y.

the virtual power theorem reads

$$\begin{cases} \mathcal{V}^{\flat}_{\text{int}}(\overline{\mathcal{P}}; \chi) = \mathcal{V}^{\flat}_{\text{ext}}(\overline{\mathcal{P}}; \chi), \\ \mathcal{V}^{\sharp}_{\text{int}}(\overline{\mathcal{P}}; \boldsymbol{\chi}) = \mathcal{V}^{\sharp}_{\text{ext}}(\overline{\mathcal{P}}; \boldsymbol{\chi}), \end{cases} \tag{1}$$

where $\mathcal{V}_{\text{int}}$ and $\mathcal{V}_{\text{ext}}$ are, respectively, the internal and external virtual power and given by

$$\begin{cases} \mathcal{V}^{\flat}_{\text{int}}(\overline{\mathcal{P}}; \chi) = \displaystyle\int_{\overline{\mathcal{P}}} (-\pi\chi + \boldsymbol{\xi}\cdot \operatorname{grad}\chi)\, \mathrm{d}v, \qquad \mathcal{V}^{\flat}_{\text{ext}}(\overline{\mathcal{P}}; \chi) = \displaystyle\int_{\overline{\mathcal{P}}} \gamma\chi\, \mathrm{d}v + \int_{\partial\overline{\mathcal{P}}} \xi_{\mathcal{S}}\chi\, \mathrm{d}a, \\ \mathcal{V}^{\sharp}_{\text{int}}(\overline{\mathcal{P}}; \boldsymbol{\chi}) = \displaystyle\int_{\overline{\mathcal{P}}} \mathbf{T} : \operatorname{grad}\boldsymbol{\chi}\, \mathrm{d}v, \qquad \mathcal{V}^{\sharp}_{\text{ext}}(\overline{\mathcal{P}}; \boldsymbol{\chi}) = \displaystyle\int_{\overline{\mathcal{P}}} \boldsymbol{b}\cdot\boldsymbol{\chi}\, \mathrm{d}v + \int_{\partial\overline{\mathcal{P}}} \boldsymbol{t}_{\mathcal{S}}\cdot\boldsymbol{\chi}\, \mathrm{d}a, \end{cases} \tag{2}$$

for any 'arbitrary' part (control volume) $\overline{\mathcal{P}}$ and admissible virtual fields χ and $\boldsymbol{\chi}$ living in a test-kinematic space $\mathcal{Y}$ and $\mathcal{Z}$, respectively. Here, integrating by parts the principle of virtual powers (1) and using the divergence theorem, we have from variational arguments that $\xi_{\mathcal{S}} := \boldsymbol{\xi}\cdot\boldsymbol{n}$ and $\boldsymbol{t}_{\mathcal{S}} := \mathbf{T}[\boldsymbol{n}]$. I remark that arbitrary in this scenario is not really arbitrary. One is sovereign to choose only arbitrary parts $\overline{\mathcal{P}}$ with a continuous outward unit normal field. Otherwise, additional interactions beyond those described in (2) should be considered.

Classically, this principle of virtual powers is a weak form where the primary objects are the linear spaces $\mathcal{Y}$ and $\mathcal{Z}$ of test-kinematic (more customary, virtual) fields. Accompanying the notion of external virtual power expended by virtual kinematic processes, χ and $\boldsymbol{\chi}$, the set of surface microtractions $\xi_{\mathcal{S}}$ and surface tractions $\boldsymbol{t}_{\mathcal{S}}$ enter as elements of the dual of $\mathcal{Y}$ and $\mathcal{Z}$, respectively. Analogously, the external microforce γ and force $\boldsymbol{b}$ enter as elements of the dual of $\mathcal{Y}$ and $\mathcal{Z}$, respectively. Conversely, accompanying the notion of internal virtual power, the set of microstress $\boldsymbol{\xi}$ and stress $\mathbf{T}$ enter as the dual of the collection of test-kinematic gradients, $\operatorname{grad}\chi$ and $\operatorname{grad}\boldsymbol{\chi}$. Also, the internal microforce π enters as an element of the dual of $\mathcal{Y}$. Podio-Guidugli & Vianello [48] regard these two duality relations as the fundamental relations of any constitutive theory. The interested reader in this formalism is referred to [48].

Synopsis, chapter 1

Microkinetics

(i) Fields:

(Primitive quantity)

αth microstress $\boldsymbol{\xi}^\alpha := \boldsymbol{\xi}^\alpha(\boldsymbol{x}, t)$

(To be derived)

αth surface microtraction $\xi_{\mathcal{S}}^\alpha := \xi_{\mathcal{S}}^\alpha(\boldsymbol{x}, t; \boldsymbol{n}, \mathbf{K}, \mathbb{K})$

αth junction-edge microtraction $\xi_{\mathcal{C}}^\alpha := \xi_{\mathcal{C}}^\alpha(\boldsymbol{x}, t; \boldsymbol{n}^\pm, \mathbf{K}^\pm)$

αth junction-point microforce $\xi_{\mathcal{O}}^\alpha := \xi_{\mathcal{O}}^\alpha(\boldsymbol{x}, t; {}^\pm\boldsymbol{n}^\pm)$

(ii) Hyperfields:

(Primitive quantities)

αth hypermicrostress $\boldsymbol{\Xi}^\alpha := \boldsymbol{\Xi}^\alpha(\boldsymbol{x}, t)$

αth supra hypermicrostress $\mathbb{X}^\alpha := \mathbb{X}^\alpha(\boldsymbol{x}, t)$

(To be derived)

αth surface hypermicrotraction ${}^2\xi_{\mathcal{S}}^\alpha := {}^2\xi_{\mathcal{S}}^\alpha(\boldsymbol{x}, t; \boldsymbol{n}, \mathbf{K})$

αth surface supra-hypermicrotraction ${}^3\xi_{\mathcal{S}}^\alpha := {}^3\xi_{\mathcal{S}}^\alpha(\boldsymbol{x}, t; \boldsymbol{n})$

αth junction-edge hypermicrotraction $\{{}^2_n\xi_{\mathcal{C}}^\alpha := {}^2_n\xi_{\mathcal{C}}^\alpha(\boldsymbol{x}, t; \boldsymbol{n}^\pm), \ {}^2_\nu\xi_{\mathcal{C}}^\alpha := {}^2_\nu\xi_{\mathcal{C}}^\alpha(\boldsymbol{x}, t; \boldsymbol{n}^\pm)\}$

(iii) Couple-fields:

(To be derived)

αth surface-couple microtraction $\varpi_{\mathcal{S}}^\alpha := \varpi_{\mathcal{S}}^\alpha(\boldsymbol{x}, t; \boldsymbol{n}, \mathbf{K})$

αth junction-edge-couple microtraction $\varpi_{\mathcal{C}}^\alpha := \varpi_{\mathcal{C}}^\alpha(\boldsymbol{x}, t; \boldsymbol{n}^\pm)$

Kinetics

(i) Fields:

(Primitive quantity)

stress $\mathbf{T} := \mathbf{T}(\boldsymbol{x}, t)$

(To be derived)

surface traction $\boldsymbol{t}_{\mathcal{S}} := \boldsymbol{t}_{\mathcal{S}}(\boldsymbol{x}, t; \boldsymbol{n}, \mathbf{K}, \mathbb{K})$

junction-edge traction $\boldsymbol{t}_{\mathcal{C}} := \boldsymbol{t}_{\mathcal{C}}(\boldsymbol{x}, t; \boldsymbol{n}^{\pm}, \mathbf{K}^{\pm})$

junction-point force $\boldsymbol{t}_{\mathcal{O}} := \boldsymbol{t}_{\mathcal{O}}(\boldsymbol{x}, t; {}^{\pm}\boldsymbol{n}^{\pm})$

(ii) Hyperfields:

(Primitive quantities)

hyperstress $\mathbb{T} := \mathbb{T}(\boldsymbol{x}, t)$

supra hyperstress $\mathbf{\mathbb{T}} := \mathbf{\mathbb{T}}(\boldsymbol{x}, t)$

(To be derived)

surface hypertraction ${}^{2}\boldsymbol{t}_{\mathcal{S}} := {}^{2}\boldsymbol{t}_{\mathcal{S}}(\boldsymbol{x}, t; \boldsymbol{n}, \mathbf{K})$

surface supra-hypertraction ${}^{3}\boldsymbol{t}_{\mathcal{S}} := {}^{3}\boldsymbol{t}_{\mathcal{S}}(\boldsymbol{x}, t; \boldsymbol{n})$

junction-edge hypertraction $\{{}^{2}_{n}\boldsymbol{t}_{\mathcal{C}} := {}^{2}_{n}\boldsymbol{t}_{\mathcal{C}}(\boldsymbol{x}, t; \boldsymbol{n}^{\pm}), {}^{2}_{\nu}\boldsymbol{t}_{\mathcal{C}} := {}^{2}_{\nu}\boldsymbol{t}_{\mathcal{C}}(\boldsymbol{x}, t; \boldsymbol{n}^{\pm})\}$

(iii) Couple-fields:

(To be derived)

surface-couple traction $\boldsymbol{m}_{\mathcal{S}} := \boldsymbol{m}_{\mathcal{S}}(\boldsymbol{x}, t; \boldsymbol{n}, \mathbf{K})$

junction-edge-couple traction $\boldsymbol{m}_{\mathcal{C}} := \boldsymbol{m}_{\mathcal{C}}(\boldsymbol{x}, t; \boldsymbol{n}^{\pm})$

Chapter 2
Integro-Differential Machinery

In this chapter, we present helpful mathematical tools to be used in the remainder of this study. We derive the relevant differential relations on a body $\mathcal{B}$, a surface $\mathcal{S}$, and an edge $\mathcal{C}$. With these differential relations, we present integro-differential relations on a part $\mathcal{P}$, on smooth and nonsmooth surfaces $\mathcal{S}$, and on smooth and nonsmooth curves $\mathcal{C}$. We conclude this chapter with the transport theorem and the mass balance equation.

1 Differential Relations on $\mathcal{P} \subset \mathcal{B}$

Consider an arbitrary part $\mathcal{P}$ embedded in a region $\mathcal{B}$ of a three-dimensional point space $\mathcal{E}$. Given the coordinates τ^i $(i = 1, 2, 3)$, the ith contravariant basis $\boldsymbol{g}^i$, and the conventional partial derivative $\partial_i := \partial/\partial\tau^i$, let κ and $\boldsymbol{\kappa}$ be, respectively, smooth scalar and vector fields on $\mathcal{B}$, whereas $\mathbf{A}$ and $\mathbb{A}$ are, respectively, second- and third-order tensor fields on $\mathcal{B}$. The gradient of a scalar field κ is defined as

$$\operatorname{grad}\kappa := \partial_i\kappa\, \boldsymbol{g}^i, \tag{3}$$

whereas the gradient of a vector field $\boldsymbol{\kappa}$, or a tensor field of order greater than zero, is defined as

$$\operatorname{grad}\boldsymbol{\kappa} := \partial_i\boldsymbol{\kappa} \otimes \boldsymbol{g}^i. \tag{4}$$

The second gradient is the result of applying recursively definitions (3) and (4), where for a scalar field κ, we have that

$$\begin{aligned}\operatorname{grad}^2\kappa &:= \partial_j(\partial_i\kappa\, \boldsymbol{g}^i) \otimes \boldsymbol{g}^j,\\ &= \partial_j\partial_i\kappa\, \boldsymbol{g}^i \otimes \boldsymbol{g}^j + \partial_i\kappa\, \partial_j\boldsymbol{g}^i \otimes \boldsymbol{g}^j,\end{aligned} \tag{5}$$

L. Espath, *Mechanics and Geometry of Enriched Continua*,
https://doi.org/10.1007/978-3-031-28934-7_2

whereas for a vector field $\boldsymbol{\kappa}$, or a tensor field of order greater than zero,

$$\begin{aligned} \operatorname{grad}^2\boldsymbol{\kappa} &:= \partial_j(\partial_i\boldsymbol{\kappa}\otimes\boldsymbol{g}^i)\otimes\boldsymbol{g}^j, \\ &= \partial_j\partial_i\boldsymbol{\kappa}\otimes\boldsymbol{g}^i\otimes\boldsymbol{g}^j + \partial_i\boldsymbol{\kappa}\otimes\partial_j\boldsymbol{g}^i\otimes\boldsymbol{g}^j. \end{aligned} \tag{6}$$

Next, the divergence of a vector field $\boldsymbol{\kappa}$ is defined as

$$\operatorname{div}\boldsymbol{\kappa} := \operatorname{grad}\boldsymbol{\kappa} : (\boldsymbol{g}^j\otimes\boldsymbol{g}_j) = \partial_i\boldsymbol{\kappa}\cdot\boldsymbol{g}^i, \tag{7}$$

whereas the divergence of a second-order tensor field $\mathbf{A}$, or a tensor field of order greater than one, is defined as

$$\operatorname{div}\mathbf{A} := \operatorname{grad}\mathbf{A}[\boldsymbol{g}^j\otimes\boldsymbol{g}_j] = \partial_i\mathbf{A}[\boldsymbol{g}^i]. \tag{8}$$

Analogously to the recursions (5) and (6), the second divergence of a second-order tensor field $\mathbf{A}$ is given by

$$\begin{aligned} \operatorname{div}^2\mathbf{A} &:= \operatorname{grad}(\operatorname{grad}\mathbf{A}[\boldsymbol{g}^k\otimes\boldsymbol{g}_k])[\boldsymbol{g}^l\otimes\boldsymbol{g}_l], \\ &= \partial_j(\partial_i\mathbf{A}[\boldsymbol{g}^i])\cdot\boldsymbol{g}^j, \\ &= \partial_j\partial_i\mathbf{A}[\boldsymbol{g}^i]\cdot\boldsymbol{g}^j + \partial_i\mathbf{A}[\partial_j\boldsymbol{g}^i]\cdot\boldsymbol{g}^j, \end{aligned} \tag{9}$$

whereas for a third-order tensor field $\mathbb{A}$, or a tensor field of order greater than two, we have that

$$\begin{aligned} \operatorname{div}^2\mathbb{A} &:= \operatorname{grad}(\operatorname{grad}\mathbb{A}[\boldsymbol{g}^k\otimes\boldsymbol{g}_k])[\boldsymbol{g}^l\otimes\boldsymbol{g}_l], \\ &= \partial_j(\partial_i\mathbb{A}[\boldsymbol{g}^i])[\boldsymbol{g}^j], \\ &= \partial_j\partial_i\mathbb{A}[\boldsymbol{g}^i][\boldsymbol{g}^j] + \partial_i\mathbb{A}[\partial_j\boldsymbol{g}^i][\boldsymbol{g}^j]. \end{aligned} \tag{10}$$

2 Differential Relations on $\mathcal{S}\subset\mathcal{P}$

Consider a smooth surface $\mathcal{S}\subset\mathcal{P}$ oriented by the unit normal $\boldsymbol{n}$ at $\boldsymbol{x}\in\mathcal{S}$. Let $\mathcal{S}$ be parameterized by coordinates τ^p with $p = 1, 2$ and z be a smooth extension of $\mathcal{S}$ along its normal $\boldsymbol{n}$ at $\boldsymbol{x}$,

$$z(\boldsymbol{x},\tau) := \boldsymbol{x} + \tau\boldsymbol{n}(\boldsymbol{x}), \qquad \forall\boldsymbol{x}\in\mathcal{S}, \tag{11}$$

with τ representing the normal coordinate n and taking values in an open interval of zero so that there exists a one-to-one mapping $z \leftrightarrow (\boldsymbol{x},\tau)$. This parameterization induces the following local covariant basis

$$\boldsymbol{g}_p := \partial_p z = \partial_p\boldsymbol{x} + \tau\partial_p\boldsymbol{n} \qquad \text{and} \qquad \boldsymbol{g}_n := \partial_n z = \boldsymbol{n}. \tag{12}$$

In view of the representation (12), at $\boldsymbol{x} = \boldsymbol{z}(\boldsymbol{x}, 0)$, we define

$$\boldsymbol{e}^p := \boldsymbol{g}^p|_{\tau=0}. \tag{13}$$

Thus,

$$\boldsymbol{e}_p = \partial_p \boldsymbol{x}. \tag{14}$$

Furthermore, the contravariant $\boldsymbol{g}^p$ and covariant $\boldsymbol{g}_q$ bases satisfy[1]

$$\boldsymbol{g}^p \cdot \boldsymbol{g}_q = \delta^p_{\cdot q}. \tag{16}$$

By taking the partial derivative of relations (12) and (16) with respect to τ, we have, respectively, that

$$\partial_n \boldsymbol{g}_p = \partial_p \boldsymbol{n}, \tag{17}$$

and

$$\partial_n \boldsymbol{g}^p \cdot \boldsymbol{g}_q = -\boldsymbol{g}^p \cdot \partial_n \boldsymbol{g}_q. \tag{18}$$

Combining the two identities above while multiplying by $\boldsymbol{g}^q$, we arrive at the identity

$$\partial_n \boldsymbol{g}^p = -(\boldsymbol{g}^q \otimes \partial_q \boldsymbol{n})[\boldsymbol{g}^p]. \tag{19}$$

In view of the reciprocal pair of bases induced by the parameterization (11), consider the differential operators as follows. The definition of the gradient of a scalar field κ at $\boldsymbol{x} \in \mathcal{S}$, given in expression (3), takes the form

$$\operatorname{grad} \kappa = \partial_n \kappa \, \boldsymbol{n} + \partial_p \kappa \, \boldsymbol{e}^p, \tag{20}$$

where

$$\partial_n \kappa = \operatorname{grad} \kappa \cdot \boldsymbol{n}, \tag{21}$$

whereas the definition of the gradient of a vector field $\boldsymbol{\kappa}$, or a tensor of order greater than zero, at $\boldsymbol{x} \in \mathcal{S}$, given in expression (4), becomes

$$\operatorname{grad} \boldsymbol{\kappa} = \partial_n \boldsymbol{\kappa} \otimes \boldsymbol{n} + \partial_p \boldsymbol{\kappa} \otimes \boldsymbol{e}^p, \tag{22}$$

where

---

[1] With an arithmetic modulo two for the index p, we may set the contravariant bases at z as

$$\boldsymbol{g}^p := (-1)^p \mathbf{N} \boldsymbol{g}_{p+1}, \qquad \text{no sum on } p, \tag{15}$$

where $\mathbf{N} := -\boldsymbol{g}^1 \otimes \boldsymbol{g}^2 + \boldsymbol{g}^2 \otimes \boldsymbol{g}^1$. Moreover, with $m = \boldsymbol{g}^1 \times \boldsymbol{g}^2 \cdot \boldsymbol{n}$, we have the volume form $\mathbf{N}_m := m\mathbf{N}$, a skew-symmetry tensor associated with the unit normal $\boldsymbol{n}$. Following the same reasoning, we may define the contravariant bases at $\boldsymbol{x}$. For further details, the interested reader is referred to [44].

$$\partial_n \boldsymbol{\kappa} = (\operatorname{grad} \boldsymbol{\kappa})[\boldsymbol{n}]. \tag{23}$$

Bearing in mind the representations (20) and (22) along with the parameterization (11), $\partial_n \boldsymbol{n}$ vanishes and the second gradient of a scalar field κ at $\boldsymbol{x} \in \mathcal{S}$, given in expression (5), is

$$\begin{aligned}\operatorname{grad}^2 \kappa = {} & \partial_q(\partial_p \kappa\, \boldsymbol{e}^p) \otimes \boldsymbol{e}^q + \partial_n \partial_p \kappa\, \boldsymbol{e}^p \otimes \boldsymbol{n} + \partial_q \partial_n \kappa\, \boldsymbol{n} \otimes \boldsymbol{e}^q \\ & + \partial_n^2 \kappa\, \boldsymbol{n} \otimes \boldsymbol{n} + \partial_p \kappa\, \partial_n \boldsymbol{e}^p \otimes \boldsymbol{n} + \partial_n \kappa\, \partial_q \boldsymbol{n} \otimes \boldsymbol{e}^q,\end{aligned} \tag{24}$$

whereas the gradient of a vector field $\boldsymbol{\kappa}$, or a tensor of order greater than zero, at $\boldsymbol{x} \in \mathcal{S}$, given in expression (6), is

$$\begin{aligned}\operatorname{grad}^2 \boldsymbol{\kappa} = {} & \partial_q(\partial_p \boldsymbol{\kappa} \otimes \boldsymbol{e}^p) \otimes \boldsymbol{e}^q + \partial_n \partial_p \boldsymbol{\kappa} \otimes \boldsymbol{e}^p \otimes \boldsymbol{n} + \partial_q \partial_n \boldsymbol{\kappa} \otimes \boldsymbol{n} \otimes \boldsymbol{e}^q \\ & + \partial_n^2 \boldsymbol{\kappa} \otimes \boldsymbol{n} \otimes \boldsymbol{n} + \partial_p \boldsymbol{\kappa} \otimes \partial_n \boldsymbol{e}^p \otimes \boldsymbol{n} + \partial_n \boldsymbol{\kappa} \otimes \partial_q \boldsymbol{n} \otimes \boldsymbol{e}^q.\end{aligned} \tag{25}$$

Similarly, the divergence (7) of a vector field $\boldsymbol{\kappa}$ at $\boldsymbol{x} \in \mathcal{S}$ takes the form

$$\operatorname{div} \boldsymbol{\kappa} = \partial_n \boldsymbol{\kappa} \cdot \boldsymbol{n} + \partial_p \boldsymbol{\kappa} \cdot \boldsymbol{e}^p, \tag{26}$$

whereas the divergence (8) of a second-order tensor field $\mathbf{A}$, or a tensor field of order greater than one, becomes

$$\operatorname{div} \mathbf{A} = \partial_n \mathbf{A}[\boldsymbol{n}] + \partial_p \mathbf{A}[\boldsymbol{e}^p]. \tag{27}$$

Analogously to the recursions (5) and (6), the second divergence of a second-order tensor field $\mathbf{A}$ is given by

$$\begin{aligned}\operatorname{div}^2 \mathbf{A} = {} & \partial_q(\partial_p \mathbf{A}[\boldsymbol{e}^p]) \cdot \boldsymbol{e}^q + \partial_n \partial_p \mathbf{A}[\boldsymbol{e}^p] \cdot \boldsymbol{n} + \partial_q \partial_n \mathbf{A}[\boldsymbol{n}] \cdot \boldsymbol{e}^q \\ & + \partial_n^2 \mathbf{A}[\boldsymbol{n}] \cdot \boldsymbol{n} + \partial_p \mathbf{A}[\partial_n \boldsymbol{e}^p] \cdot \boldsymbol{n} + \partial_n \mathbf{A}[\partial_q \boldsymbol{n}] \cdot \boldsymbol{e}^q,\end{aligned} \tag{28}$$

while for a third-order tensor field $\mathbb{A}$, or a tensor field of order greater than two, we have that

$$\begin{aligned}\operatorname{div}^2 \mathbb{A} = {} & \partial_q(\partial_p \mathbb{A}[\boldsymbol{e}^p])[\boldsymbol{e}^q] + \partial_n \partial_p \mathbb{A}[\boldsymbol{e}^p][\boldsymbol{n}] + \partial_q \partial_n \mathbb{A}[\boldsymbol{n}][\boldsymbol{e}^q] \\ & + \partial_n^2 \mathbb{A}[\boldsymbol{n}][\boldsymbol{n}] + \partial_p \mathbb{A}[\partial_n \boldsymbol{e}^p][\boldsymbol{n}] + \partial_n \mathbb{A}[\partial_q \boldsymbol{n}][\boldsymbol{e}^q].\end{aligned} \tag{29}$$

Again, note that in expressions (24), (25), (28), and (29), given the parameterization (11), we used, respectively, $\partial_n \kappa\, \partial_n \boldsymbol{n} \otimes \boldsymbol{n} = \mathbf{0}$, $\partial_n \boldsymbol{\kappa} \otimes \partial_n \boldsymbol{n} \otimes \boldsymbol{n} = \mathbf{0}$, $\partial_n \mathbf{A}[\partial_n \boldsymbol{n}] \cdot \boldsymbol{n} = 0$, and $\partial_n \mathbb{A}[\partial_n \boldsymbol{n}][\boldsymbol{n}] = \mathbf{0}$.

Next, let $\mathbf{P}_n := \mathbf{P}_n(\boldsymbol{n})$[2] denote the projector onto the plane defined by $\boldsymbol{n}$ at $\boldsymbol{x} \in \mathcal{S}$, which reads

[2] The second-order tensor $\mathbf{P}_\alpha$, with $\boldsymbol{\alpha}$ being a vector, is defined by $\mathbf{P}_\alpha := \mathbf{1} - \boldsymbol{\alpha} \otimes \boldsymbol{\alpha}$. Also, $\mathbf{P}_{\alpha\beta} := \mathbf{1} - \boldsymbol{\alpha} \otimes \boldsymbol{\alpha} - \boldsymbol{\beta} \otimes \boldsymbol{\beta}$.

$$\mathbf{P}_n := \mathbf{1} - \boldsymbol{n} \otimes \boldsymbol{n} = \mathbf{P}_n^\top. \tag{30}$$

In view of the expressions (20) and (22) along with (30), let the surface gradients of a scalar field κ and a vector field $\boldsymbol{\kappa}$, or a tensor field of order greater than zero, be, respectively, represented by

$$\begin{cases} \operatorname{grad}_{\mathcal{S}}\kappa := \partial_p \kappa\, \boldsymbol{e}^p = \mathbf{P}_n[\operatorname{grad}\kappa], \\ \operatorname{grad}_{\mathcal{S}}\boldsymbol{\kappa} := \partial_p \boldsymbol{\kappa} \otimes \boldsymbol{e}^p = (\operatorname{grad}\boldsymbol{\kappa})\mathbf{P}_n. \end{cases} \tag{31}$$

With expressions (31), we can additively decompose the gradient into the normal and tangential components. Here, the normal components are related to normal derivatives whereas tangential components are related to surface gradients. Thus, from expressions (20) and (22) in conjunction with (31), we are led to

$$\operatorname{grad}\kappa = (\operatorname{grad}\kappa)^{\text{nor}} + (\operatorname{grad}\kappa)^{\text{tan}}, \tag{32}$$

with

$$\begin{cases} (\operatorname{grad}\kappa)^{\text{nor}} := \partial_n \kappa\, \boldsymbol{n}, \\ (\operatorname{grad}\kappa)^{\text{tan}} := \operatorname{grad}_{\mathcal{S}}\kappa, \end{cases} \tag{33}$$

and

$$\operatorname{grad}\boldsymbol{\kappa} = (\operatorname{grad}\boldsymbol{\kappa})^{\text{nor}} + (\operatorname{grad}\boldsymbol{\kappa})^{\text{tan}}, \tag{34}$$

with

$$\begin{cases} (\operatorname{grad}\boldsymbol{\kappa})^{\text{nor}} := \partial_n \boldsymbol{\kappa} \otimes \boldsymbol{n}, \\ (\operatorname{grad}\boldsymbol{\kappa})^{\text{tan}} := \operatorname{grad}_{\mathcal{S}}\boldsymbol{\kappa}. \end{cases} \tag{35}$$

Additionally, given the surface gradient definition of a vector field (31), we are led to define the curvature tensor, the negative surface gradient of the unit normal, as

$$\mathbf{K} := -\partial_p \boldsymbol{n} \otimes \boldsymbol{e}^p = -\operatorname{grad}_{\mathcal{S}}\boldsymbol{n}. \tag{36}$$

Also, let the surface gradient of the curvature tensor (36) be

$$\mathbb{K} := -\partial_q(\partial_p \boldsymbol{n} \otimes \boldsymbol{e}^p) \otimes \boldsymbol{e}^q = -\operatorname{grad}^2_{\mathcal{S}}\boldsymbol{n} = \operatorname{grad}_{\mathcal{S}}\mathbf{K}. \tag{37}$$

We say that $\mathbb{K}$ is the second curvature tensor.

Next, in view of the expressions (24) and (25), let the second surface gradients of a scalar field κ and a vector field $\boldsymbol{\kappa}$, or a tensor field of order greater than zero, be, respectively,

$$\begin{cases} \operatorname{grad}^2_{\mathcal{S}}\kappa := \partial_q(\partial_p \kappa\, \boldsymbol{e}^p) \otimes \boldsymbol{e}^q = (\operatorname{grad}(\mathbf{P}_n[\operatorname{grad}\kappa]))\mathbf{P}_n, \\ \operatorname{grad}^2_{\mathcal{S}}\boldsymbol{\kappa} := \partial_q(\partial_p \boldsymbol{\kappa} \otimes \boldsymbol{e}^p) \otimes \boldsymbol{e}^q = (\operatorname{grad}((\operatorname{grad}\boldsymbol{\kappa})\mathbf{P}_n))\mathbf{P}_n. \end{cases} \tag{38}$$

Similarly, in view of (26) and (27), the surface divergence of a vector field $\boldsymbol{\kappa}$ and a tensor field $\mathbf{A}$, or a tensor field of order greater than one, are, respectively,

$$\begin{cases} \operatorname{div}_S \boldsymbol{\kappa} = \partial_p \boldsymbol{\kappa} \cdot \boldsymbol{e}^p = \operatorname{grad} \boldsymbol{\kappa} : \mathbf{P}_n, \\ \operatorname{div}_S \mathbf{A} = \partial_p \mathbf{A}[\boldsymbol{e}^p] = (\operatorname{grad} \mathbf{A})[\mathbf{P}_n]. \end{cases} \tag{39}$$

Based on the surface divergence $(39)_1$ and the curvature tensor (36), we can also define the mean curvature by

$$K := \tfrac{1}{2} \operatorname{tr} \mathbf{K} = -\tfrac{1}{2} \operatorname{div}_S \boldsymbol{n}. \tag{40}$$

In view of (28) and (29), the second surface divergence of a second-order tensor field $\mathbf{A}$ and a third-order tensor field $\mathbb{A}$, or a tensor field of order greater than two, are, respectively,

$$\begin{cases} \operatorname{div}_S^2 \mathbf{A} = \partial_q (\partial_p \mathbf{A}[\boldsymbol{e}^p]) \cdot \boldsymbol{e}^q = \operatorname{grad}((\operatorname{grad} \mathbf{A})[\mathbf{P}_n]) : \mathbf{P}_n, \\ \operatorname{div}_S^2 \mathbb{A} = \partial_q (\partial_p \mathbb{A}[\boldsymbol{e}^p])[\boldsymbol{e}^q] = (\operatorname{grad}((\operatorname{grad} \mathbb{A})[\mathbf{P}_n]))[\mathbf{P}_n]. \end{cases} \tag{41}$$

Some immediate properties related to the curvature tensor and the second curvature tensor are as follows. First, by multiplying the curvature tensor (36) by the normal $\boldsymbol{n}$, we arrive at

$$\mathbf{K}[\boldsymbol{n}] = \mathbf{0}. \tag{42}$$

It is also known that the curvature tensor is symmetric, that is,

$$\mathbf{K} = \mathbf{K}^{\top}. \tag{43}$$

Second, by multiplying the second curvature tensor (37) by the normal $\boldsymbol{n}$, we have that

$$\mathbb{K}[\boldsymbol{n}] = \mathbf{0}. \tag{44}$$

Third, by computing $\operatorname{grad}_S(\mathbf{K}[\boldsymbol{n}]) = \mathbf{0}$, we arrive at

$$\mathbb{K}^{\top}[\boldsymbol{n}] = \mathbf{K}^2, \tag{45}$$

and given the symmetry of the curvature tensor $\mathbf{K}$, the second curvature tensor $\mathbb{K}$ is symmetric under the permutation of its first two arguments, that is,

$$\mathbb{K} = \mathbb{K}^{\perp}. \tag{46}$$

The Cayley–Hamilton theorem for the curvature tensor $\mathbf{K}$, given by Biria et al. [49], is given by $\mathbf{K}^2 - I_1(\mathbf{K})\mathbf{K} + I_2(\mathbf{K})\mathbf{P}_n = \mathbf{0}$ with $I_1(\mathbf{K})$ and $I_2(\mathbf{K})$ being the first two invariants of the curvature tensor $\mathbf{K}$. The first invariant is $I_1(\mathbf{K}) = \frac{1}{2} \operatorname{tr} \mathbf{K} = K$ while the second invariant is $I_2(\mathbf{K}) = \frac{1}{2}((\operatorname{tr}\mathbf{K})^2 - \operatorname{tr}(\mathbf{K}^2)) = G$ being G the Gaussian curvature. Then, we may write

$$\mathbf{K}^2 - 2K\mathbf{K} + G\mathbf{P}_n = \mathbf{0}, \tag{47}$$

and express property (45) in the form

$$\mathbb{K}^{\top}[\boldsymbol{n}] = 2K\mathbf{K} - G\mathbf{P}_n. \tag{48}$$

By multiplying (45) and (48) by $\boldsymbol{n}$, we have that

$$\mathbb{K}[\boldsymbol{n}][\boldsymbol{n}] = \boldsymbol{n}\cdot\mathbb{K}^{\top}[\boldsymbol{n}] = \mathbb{K}^{\top}[\boldsymbol{n}][\boldsymbol{n}] = \mathbf{0}. \tag{49}$$

Also, for an element $\boldsymbol{x}$ of a three-dimensional point space $\mathcal{E}$ and an arbitrary but fixed origin $\boldsymbol{o}$, we have that $\operatorname{grad}\boldsymbol{r} = \mathbf{1}$ with $\boldsymbol{r} := \boldsymbol{x} - \boldsymbol{o}$. Thus,

$$\operatorname{grad}_{\mathcal{S}}\boldsymbol{r} = \mathbf{P}_n \quad \text{and} \quad \operatorname{div}_{\mathcal{S}}\boldsymbol{r} = 2. \tag{50}$$

Additionally, with (19) evaluated at $\tau = 0$ and the definition of the curvature tensor $\mathbf{K}$ (36), we have that the penultimate term in expression (24) becomes

$$\begin{aligned}
\partial_p\kappa\,\partial_n\boldsymbol{e}^p\otimes\boldsymbol{n} &= -\partial_p\kappa(\boldsymbol{e}^q\otimes\partial_q\boldsymbol{n})[\boldsymbol{e}^p]\otimes\boldsymbol{n},\\
&= -\partial_p\kappa\,\boldsymbol{e}^p\cdot(\partial_q\boldsymbol{n}\otimes\boldsymbol{e}^q)\otimes\boldsymbol{n},\\
&= (\operatorname{grad}_{\mathcal{S}}\kappa)\cdot\mathbf{K}\otimes\boldsymbol{n}.
\end{aligned} \tag{51}$$

Similarly, we have that the penultimate term in expression (25) reads

$$\begin{aligned}
\partial_p\boldsymbol{\kappa}\otimes\partial_n\boldsymbol{e}^p\otimes\boldsymbol{n} &= -\partial_p\boldsymbol{\kappa}\otimes(\boldsymbol{e}^q\otimes\partial_q\boldsymbol{n})[\boldsymbol{e}^p]\otimes\boldsymbol{n},\\
&= -(\partial_p\boldsymbol{\kappa}\otimes\boldsymbol{e}^p)(\partial_q\boldsymbol{n}\otimes\boldsymbol{e}^q)\otimes\boldsymbol{n},\\
&= (\operatorname{grad}_{\mathcal{S}}\boldsymbol{\kappa})\mathbf{K}\otimes\boldsymbol{n}.
\end{aligned} \tag{52}$$

In view of the identities above,[3] the second gradients given in expressions (24) and (25) may be rewritten as

$$\begin{aligned}
\operatorname{grad}^2\kappa = \operatorname{grad}^2_{\mathcal{S}}\kappa + \partial_n\partial_p\kappa\,(\boldsymbol{e}^p\otimes\boldsymbol{n} + \boldsymbol{n}\otimes\boldsymbol{e}^p) + \partial_n^2\kappa\,\boldsymbol{n}\otimes\boldsymbol{n}\\
+ (\operatorname{grad}_{\mathcal{S}}\kappa)\cdot\mathbf{K}\otimes\boldsymbol{n} - \partial_n\kappa\,\mathbf{K},
\end{aligned} \tag{53}$$

and

[3] Identities (19), (51), and (52) originally appeared in the works by Carrillo et al. [44] and are of utmost importance for this study.

$$\operatorname{grad}^2\boldsymbol{\kappa} = \operatorname{grad}_S^2\boldsymbol{\kappa} + \partial_n\partial_p\boldsymbol{\kappa}\otimes(\boldsymbol{e}^p\otimes\boldsymbol{n}+\boldsymbol{n}\otimes\boldsymbol{e}^p) + \partial_n^2\boldsymbol{\kappa}\otimes\boldsymbol{n}\otimes\boldsymbol{n}$$
$$+ (\operatorname{grad}_S\boldsymbol{\kappa})\mathbf{K}\otimes\boldsymbol{n} - \partial_n\boldsymbol{\kappa}\otimes\mathbf{K}. \tag{54}$$

As we decomposed expressions (32) and (34), we decompose additively the second gradient into the normal and tangential components, that is,

$$\operatorname{grad}^2\kappa = (\operatorname{grad}^2\kappa)^{\mathrm{nor}} + (\operatorname{grad}^2\kappa)^{\mathrm{tan}}, \tag{55}$$

with

$$\begin{cases} (\operatorname{grad}^2\kappa)^{\mathrm{nor}} := \partial_n\partial_p\kappa\,(\boldsymbol{e}^p\otimes\boldsymbol{n}+\boldsymbol{n}\otimes\boldsymbol{e}^p) + \partial_n^2\kappa\,\boldsymbol{n}\otimes\boldsymbol{n} - \partial_n\kappa\,\mathbf{K}, \\ (\operatorname{grad}^2\kappa)^{\mathrm{tan}} := \operatorname{grad}_S^2\kappa + (\operatorname{grad}_S\kappa)\cdot\mathbf{K}\otimes\boldsymbol{n}, \end{cases} \tag{56}$$

and

$$\operatorname{grad}^2\boldsymbol{\kappa} = (\operatorname{grad}^2\boldsymbol{\kappa})^{\mathrm{nor}} + (\operatorname{grad}^2\boldsymbol{\kappa})^{\mathrm{tan}}, \tag{57}$$

with

$$\begin{cases} (\operatorname{grad}^2\boldsymbol{\kappa})^{\mathrm{nor}} := \partial_n\partial_p\boldsymbol{\kappa}\otimes(\boldsymbol{e}^p\otimes\boldsymbol{n}+\boldsymbol{n}\otimes\boldsymbol{e}^p) + \partial_n^2\boldsymbol{\kappa}\otimes\boldsymbol{n}\otimes\boldsymbol{n} - \partial_n\boldsymbol{\kappa}\otimes\mathbf{K}, \\ (\operatorname{grad}^2\boldsymbol{\kappa})^{\mathrm{tan}} := \operatorname{grad}_S^2\boldsymbol{\kappa} + (\operatorname{grad}_S\boldsymbol{\kappa})\mathbf{K}\otimes\boldsymbol{n}. \end{cases} \tag{58}$$

Next, consider, respectively, the following identities for a vector field $\boldsymbol{\kappa}$, and a second-order tensor field $\mathbf{A}$, or a tensor field of order greater than two,

$$\operatorname{div}_S\boldsymbol{\kappa} = \operatorname{div}_S(\mathbf{P}_n[\boldsymbol{\kappa}]) + \operatorname{div}_S((\boldsymbol{\kappa}\cdot\boldsymbol{n})\boldsymbol{n}) = \operatorname{div}_S(\mathbf{P}_n[\boldsymbol{\kappa}]) - 2K\boldsymbol{\kappa}\cdot\boldsymbol{n}, \tag{59}$$

and

$$\operatorname{div}_S\mathbf{A} = \operatorname{div}_S(\mathbf{A}\mathbf{P}_n) + \operatorname{div}_S(\mathbf{A}[\boldsymbol{n}]\otimes\boldsymbol{n}) = \operatorname{div}_S(\mathbf{A}\mathbf{P}_n) - 2K\mathbf{A}[\boldsymbol{n}]. \tag{60}$$

In view of the identities (59) and (60), with the particular vector $\mathbf{A}[\boldsymbol{n}]$ and the second-order tensor $\mathbb{A}[\boldsymbol{n}]$, the divergence may be decomposed, respectively, as

$$\operatorname{div}_S(\mathbf{A}[\boldsymbol{n}]) = (\operatorname{div}\mathbf{A}^\top)\cdot\boldsymbol{n} - (\operatorname{grad}\mathbf{A}) \,\vdots\, (\boldsymbol{n}\otimes\boldsymbol{n}\otimes\boldsymbol{n}) - \mathbf{A}:\mathbf{K}, \tag{61}$$

and

$$\operatorname{div}_S(\mathbb{A}[\boldsymbol{n}]) = (\operatorname{div}\mathbb{A}^\top)[\boldsymbol{n}] - (\operatorname{grad}\mathbb{A})[\boldsymbol{n}\otimes\boldsymbol{n}\otimes\boldsymbol{n}] - \mathbb{A}[\mathbf{K}]. \tag{62}$$

These are particularly useful when aiming at classical representations for surface tractions and surface microtractions.

3 Differential Relations on $\mathcal{C} \subset \mathcal{S}$

Consider a smooth curve $\mathcal{C} \subset \mathcal{S}$ oriented by the unit tangent $\boldsymbol{\sigma}$ at $\boldsymbol{x} \in \mathcal{C}$. Let $\mathcal{C}$ be parameterized by the coordinate σ. Quantities living on a curve $\mathcal{C} \subset \mathcal{S}$ may be characterized by a Darboux frame, composed of the unit tangent $\boldsymbol{\sigma}$, the unit normal $\boldsymbol{n}$ of $\mathcal{S}$, and the outward unit tangent-normal $\boldsymbol{\nu} := \boldsymbol{\sigma} \times \boldsymbol{n}$, at $\boldsymbol{x} \in \mathcal{C}$. In the Darboux frame, a vector field $\boldsymbol{\kappa}$ is expressed as

$$\boldsymbol{\kappa} = (\boldsymbol{\kappa} \cdot \boldsymbol{\sigma})\boldsymbol{\sigma} + (\boldsymbol{\kappa} \cdot \boldsymbol{n})\boldsymbol{n} + (\boldsymbol{\kappa} \cdot \boldsymbol{\nu})\boldsymbol{\nu}. \tag{63}$$

while the components of the gradient of a smooth scalar field κ and vector field $\boldsymbol{\kappa}$, in this frame, read, respectively,

$$\partial_\sigma \kappa = (\operatorname{grad} \kappa) \cdot \boldsymbol{\sigma}, \qquad \partial_n \kappa = (\operatorname{grad} \kappa) \cdot \boldsymbol{n}, \qquad \text{and} \qquad \partial_\nu \kappa = (\operatorname{grad} \kappa) \cdot \boldsymbol{\nu}, \tag{64}$$

and

$$\partial_\sigma \boldsymbol{\kappa} = (\operatorname{grad} \boldsymbol{\kappa})[\boldsymbol{\sigma}], \qquad \partial_n \boldsymbol{\kappa} = (\operatorname{grad} \boldsymbol{\kappa})[\boldsymbol{n}], \qquad \text{and} \qquad \partial_\nu \boldsymbol{\kappa} = (\operatorname{grad} \boldsymbol{\kappa})[\boldsymbol{\nu}]. \tag{65}$$

On a curve $\mathcal{C}$, the Darboux frame $\{\boldsymbol{\sigma}, \boldsymbol{n}, \boldsymbol{\nu}\}$ obeys the following relations

$$\begin{cases} \partial_\sigma \boldsymbol{\sigma} = k_n \boldsymbol{n} - k_g \boldsymbol{\nu}, \\ \partial_\sigma \boldsymbol{n} = -k_n \boldsymbol{\sigma} - t_g \boldsymbol{\nu}, \\ \partial_\sigma \boldsymbol{\nu} = k_g \boldsymbol{\sigma} + t_g \boldsymbol{n}, \end{cases} \tag{66}$$

where

$$\begin{cases} k_n = \partial_\sigma \boldsymbol{\sigma} \cdot \boldsymbol{n} = -\partial_\sigma \boldsymbol{n} \cdot \boldsymbol{\sigma}, \\ k_g = \partial_\sigma \boldsymbol{\nu} \cdot \boldsymbol{\sigma} = -\partial_\sigma \boldsymbol{\sigma} \cdot \boldsymbol{\nu}, \\ t_g = \partial_\sigma \boldsymbol{\nu} \cdot \boldsymbol{n} = -\partial_\sigma \boldsymbol{n} \cdot \boldsymbol{\nu}, \end{cases} \tag{67}$$

where k_n and k_g are the normal and geodesic curvatures of $\mathcal{C}$, while t_g is the geodesic torsion of $\mathcal{C}$. Moreover, in view of (50), we have that

$$(\operatorname{grad} \boldsymbol{r})\mathbf{P}_{n\nu} = \partial_\sigma \boldsymbol{r} \otimes \boldsymbol{\sigma} = \mathbf{P}_{n\nu}, \tag{68}$$

where $\mathbf{P}_{n\nu} := \mathbf{1} - \boldsymbol{n} \otimes \boldsymbol{n} - \boldsymbol{\nu} \otimes \boldsymbol{\nu}$.

Next, considering the following representation for the surface gradients of the scalar field κ and vector field $\boldsymbol{\kappa}$,

$$\begin{cases} \operatorname{grad}_{\mathcal{S}} \kappa = \partial_\sigma \kappa \, \boldsymbol{\sigma} + \partial_\nu \kappa \, \boldsymbol{\nu}, \\ \operatorname{grad}_{\mathcal{S}} \boldsymbol{\kappa} = \partial_\sigma \boldsymbol{\kappa} \otimes \boldsymbol{\sigma} + \partial_\nu \boldsymbol{\kappa} \otimes \boldsymbol{\nu}, \end{cases} \tag{69}$$

we obtain the following identities,

$$\boldsymbol{a} \cdot \mathrm{grad}_{\mathcal{S}} \kappa = \partial_\sigma(\kappa\, \boldsymbol{a} \cdot \boldsymbol{\sigma}) + \partial_\nu \kappa\, \boldsymbol{a} \cdot \boldsymbol{\nu} - \kappa\, \partial_\sigma(\boldsymbol{a} \cdot \boldsymbol{\sigma}), \tag{70}$$

and

$$\mathbf{A} : \mathrm{grad}_{\mathcal{S}} \boldsymbol{\kappa} = \partial_\sigma(\boldsymbol{\kappa} \cdot \mathbf{A}[\boldsymbol{\sigma}]) + \partial_\nu \boldsymbol{\kappa} \cdot \mathbf{A}[\boldsymbol{\nu}] - \boldsymbol{\kappa} \cdot \partial_\sigma(\mathbf{A}[\boldsymbol{\sigma}]). \tag{71}$$

4 Integro-Differential Relations on $\mathcal{S} \subset \mathcal{P}$

Consider a surface $\mathcal{S}$ oriented by the unit normal $\boldsymbol{n}$ at $\boldsymbol{x} \in \mathcal{S}$ as depicted in Fig. 1. The surface $\mathcal{S}$ is an open surface with a boundary-edge $\partial\mathcal{S}$ with the outward unit tangent-normal $\boldsymbol{\nu}$ at $\boldsymbol{x} \in \partial\mathcal{S}$. The boundary-edge $\partial\mathcal{S}$ is oriented by the unit tangent $\boldsymbol{\sigma} := \boldsymbol{n} \times \boldsymbol{\nu}$ at $\boldsymbol{x} \in \partial\mathcal{S}$. The surface $\mathcal{S}$ may lose its smoothness along a junction-edge $\mathcal{C}$. In a neighborhood of $\mathcal{C}$, two smooth surfaces $\mathcal{S}^{\pm}$ are defined where $\mathcal{C}$ may be characterized by the limiting normals $\{\boldsymbol{n}^+, \boldsymbol{n}^-\}$ of $\mathcal{S}^{\pm}$, or equivalently by the limiting unit tangent-normals $\{\boldsymbol{\nu}^+, \boldsymbol{\nu}^-\}$. We let $\mathcal{C}$ be oriented by the unit tangent $\boldsymbol{\sigma} := \boldsymbol{\sigma}^+$. The boundary-edge may also lose its smoothness at a junction-point on $\partial\mathcal{S}$, namely, $\partial^2\mathcal{S}$. The junction-edge $\mathcal{C}$ is an open curve with a junction-edge boundary $_{\partial\mathcal{C}}$. The points $\partial^2\mathcal{S}$ and $_{\partial\mathcal{C}}$ are identical, that is, $\partial^2\mathcal{S} \equiv {}_{\partial\mathcal{C}}$. At $\partial^2\mathcal{S}$, the unit tangent field is discontinuous with limiting values of ${}^-\boldsymbol{\sigma}$ and ${}^+\boldsymbol{\sigma}$. Similarly, the junction-edge $\mathcal{C}$ may also lose smoothness at a junction-point $_{\mathcal{C}}$, where the unit tangent takes the limiting values of ${}^-\boldsymbol{\sigma}$ and ${}^+\boldsymbol{\sigma}$.

For any smooth vector field $\boldsymbol{\kappa}$ and tensor field $\mathbf{A}$ on a smooth closed oriented surface $\mathcal{S}$, the surface divergence theorem states that

$$\begin{cases} \displaystyle\int_{\mathcal{S}} \mathrm{div}_{\mathcal{S}}(\mathbf{P}_n[\boldsymbol{\kappa}])\, \mathrm{d}a = 0, \\ \displaystyle\int_{\mathcal{S}} \mathrm{div}_{\mathcal{S}}(\mathbf{A}\mathbf{P}_n)\, \mathrm{d}a = \mathbf{0}, \end{cases} \tag{72}$$

whereas on a smooth open oriented surface $\mathcal{S}$, (72) becomes

$$\begin{cases} \displaystyle\int_{\mathcal{S}} \mathrm{div}_{\mathcal{S}}(\mathbf{P}_n[\boldsymbol{\kappa}])\, \mathrm{d}a = \int_{\partial\mathcal{S}} \boldsymbol{\kappa} \cdot \boldsymbol{\nu}\, \mathrm{d}\sigma, \\ \displaystyle\int_{\mathcal{S}} \mathrm{div}_{\mathcal{S}}(\mathbf{A}\mathbf{P}_n)\, \mathrm{d}a = \int_{\partial\mathcal{S}} \mathbf{A}[\boldsymbol{\nu}]\, \mathrm{d}\sigma. \end{cases} \tag{73}$$

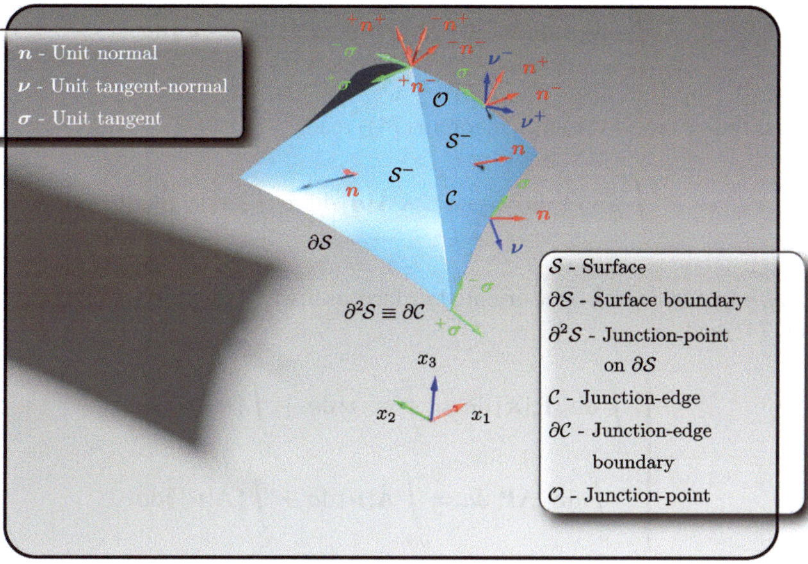

Fig. 1 Nonsmooth open surface $\mathcal{S}$ with a boundary-edge $\partial\mathcal{S}$ with the outward unit tangent-normal $\boldsymbol{\nu}$ and oriented by the unit tangent $\boldsymbol{\sigma} := \boldsymbol{n} \times \boldsymbol{\nu}$. The surface $\mathcal{S}$ lacks smoothness at a junction-edge $\mathcal{C}$, whereas the junction-edge $\mathcal{C}$ lacks smoothness at a junction-point $\mathcal{O}$. The junction-edge $\mathcal{C}$ is an open curve with a junction-edge boundary $\partial\mathcal{C}$. The boundary-edge also lacks smoothness at a junction-point on $\partial\mathcal{S}$, namely, $\partial^2\mathcal{S}$. The points $\partial^2\mathcal{S}$ and $\partial\mathcal{C}$ are identical, that is, $\partial^2\mathcal{S} \equiv \partial\mathcal{C}$. At $\partial^2\mathcal{S}$, the unit tangent field is discontinuous with limiting values of ${}^{-}\boldsymbol{\sigma}$ and ${}^{+}\boldsymbol{\sigma}$. Similarly, at a junction-point $\mathcal{O}$ the unit tangent takes the limiting values of ${}^{-}\boldsymbol{\sigma}$ and ${}^{+}\boldsymbol{\sigma}$

Here, we identify the following tangential components $\mathbf{P}_{\!n}[\boldsymbol{\kappa}]$ and $\mathbf{A}\mathbf{P}_{\!n}$ since they annihilate the unit normal $\boldsymbol{n}$.[4]

Owing to the lack of smoothness at a junction-edge $\mathcal{C}$, for any smooth vector field $\boldsymbol{\kappa}$ and tensor field $\mathbf{A}$ on a nonsmooth closed oriented surface $\mathcal{S}$ with limiting outward unit tangent-normals $\boldsymbol{\nu}^{+}$ and $\boldsymbol{\nu}^{-}$ at $\mathcal{C}$, the surface divergence theorem exhibits a *surplus*, that is,

$$\begin{cases} \displaystyle\int_{\mathcal{S}} \operatorname{div}_{\mathcal{S}}(\mathbf{P}_{\!n}[\boldsymbol{\kappa}])\,\mathrm{d}a = \int_{\mathcal{C}} \{\!\!\{\boldsymbol{\kappa}\cdot\boldsymbol{\nu}\}\!\!\}\mathrm{d}\sigma, \\ \displaystyle\int_{\mathcal{S}} \operatorname{div}_{\mathcal{S}}(\mathbf{A}\mathbf{P}_{\!n})\,\mathrm{d}a = \int_{\mathcal{C}} \{\!\!\{\mathbf{A}[\boldsymbol{\nu}]\}\!\!\}\mathrm{d}\sigma, \end{cases} \tag{74}$$

where $\{\!\!\{\boldsymbol{\kappa}\cdot\boldsymbol{\nu}\}\!\!\} := \boldsymbol{\kappa}\cdot\boldsymbol{\nu}^{+} + \boldsymbol{\kappa}\cdot\boldsymbol{\nu}^{-}$ and $\{\!\!\{\mathbf{A}[\boldsymbol{\nu}]\}\!\!\} := \mathbf{A}[\boldsymbol{\nu}^{+}] + \mathbf{A}[\boldsymbol{\nu}^{-}]$. With the identities (59) and (60), the surface divergence theorem $(74)_1$ becomes

[4] Note that, the decomposition into normal and tangential components of the gradients and second gradients in expressions (32), (34) (55), and (57) differs from our definition of tangential vectors, which are orthogonal to the unit normal, and tangential tensors, which annihilate the unit normal.

$$\int_{\mathcal{S}} \mathrm{div}_{\mathcal{S}}\boldsymbol{\kappa}\, \mathrm{d}a = -\int_{\mathcal{S}} 2K\boldsymbol{\kappa}\cdot\boldsymbol{n}\, \mathrm{d}a + \int_{\mathcal{C}} \{\!\{\boldsymbol{\kappa}\cdot\boldsymbol{\nu}\}\!\}\mathrm{d}\sigma, \tag{75}$$

whereas the surface divergence theorem $(74)_2$ reads

$$\int_{\mathcal{S}} \mathrm{div}_{\mathcal{S}}\mathbf{A}\, \mathrm{d}a = -\int_{\mathcal{S}} 2K\mathbf{A}[\boldsymbol{n}]\, \mathrm{d}a + \int_{\mathcal{C}} \{\!\{\mathbf{A}[\boldsymbol{\nu}]\}\!\}\mathrm{d}\sigma. \tag{76}$$

Finally, on a nonsmooth open oriented surface, as in Fig. 1, the surface divergence theorem (74) reads

$$\begin{cases} \displaystyle\int_{\mathcal{S}} \mathrm{div}_{\mathcal{S}}\mathbf{P}_{n}[\boldsymbol{\kappa}]\, \mathrm{d}a = \int_{\partial\mathcal{S}} \boldsymbol{\kappa}\cdot\boldsymbol{\nu}\, \mathrm{d}\sigma + \int_{\mathcal{C}} \{\!\{\boldsymbol{\kappa}\cdot\boldsymbol{\nu}\}\!\}\mathrm{d}\sigma, \\ \displaystyle\int_{\mathcal{S}} \mathrm{div}_{\mathcal{S}}\mathbf{A}\mathbf{P}_{n}\, \mathrm{d}a = \int_{\partial\mathcal{S}} \mathbf{A}[\boldsymbol{\nu}]\, \mathrm{d}\sigma + \int_{\mathcal{C}} \{\!\{\mathbf{A}[\boldsymbol{\nu}]\}\!\}\mathrm{d}\sigma. \end{cases} \tag{77}$$

With the identities (59) and (60), the surface divergence theorem $(77)_1$ becomes

$$\int_{\mathcal{S}} \mathrm{div}_{\mathcal{S}}\boldsymbol{\kappa}\, \mathrm{d}a = -\int_{\mathcal{S}} 2K\boldsymbol{\kappa}\cdot\boldsymbol{n}\, \mathrm{d}a + \int_{\partial\mathcal{S}} \boldsymbol{\kappa}\cdot\boldsymbol{\nu}\, \mathrm{d}\sigma + \int_{\mathcal{C}} \{\!\{\boldsymbol{\kappa}\cdot\boldsymbol{\nu}\}\!\}\mathrm{d}\sigma, \tag{78}$$

whereas the surface divergence theorem $(77)_2$ reads

$$\int_{\mathcal{S}} \mathrm{div}_{\mathcal{S}}\mathbf{A}\, \mathrm{d}a = -\int_{\mathcal{S}} 2K\mathbf{A}[\boldsymbol{n}]\, \mathrm{d}a + \int_{\partial\mathcal{S}} \mathbf{A}[\boldsymbol{\nu}]\, \mathrm{d}\sigma + \int_{\mathcal{C}} \{\!\{\mathbf{A}[\boldsymbol{\nu}]\}\!\}\mathrm{d}\sigma. \tag{79}$$

5 Integro-Differential Relations on $\mathcal{C} \subset \mathcal{S}$

Consider a curve $\mathcal{C}$ oriented by the unit tangent $\boldsymbol{\sigma}$ at $\boldsymbol{x} \in \mathcal{C}$. If $\mathcal{C}$ is an open curve, then consider it with boundary-points $\partial\mathcal{C}|_{\sigma=0}$ and $\partial\mathcal{C}|_{\sigma=1}$. Similarly to surfaces, a boundary- and junction-edge may lose smoothness at a point, namely, junction-point $\circ$. In analyzing a neighborhood of a junction-point $\circ$, two smooth curves $\mathcal{C}^{\pm}$ are defined. The limiting unit tangents of $\mathcal{C}^{\pm}$ at $\circ$ are denoted by the pair $\{{}^{+}\boldsymbol{\sigma},\ {}^{-}\boldsymbol{\sigma}\}$, which characterizes the junction-point $\circ$.

For any smooth vector field $\boldsymbol{\kappa}$ and tensor field $\mathbf{A}$ on a smooth closed curve $\mathcal{C}$ oriented by the tangent $\boldsymbol{\sigma}$, the gradient theorem may be stated as

$$\begin{cases} \displaystyle\int_{\mathcal{C}} \partial_\sigma(\boldsymbol{\kappa}\cdot\boldsymbol{\sigma})\,\mathrm{d}\sigma = 0, \\ \displaystyle\int_{\mathcal{C}} \partial_\sigma(\mathbf{A}[\boldsymbol{\sigma}])\,\mathrm{d}\sigma = \mathbf{0}, \end{cases} \tag{80}$$

whereas on a smooth open oriented curve $\mathcal{C}$, (80) becomes

$$\begin{cases} \displaystyle\int_{\mathcal{C}} \partial_\sigma(\boldsymbol{\kappa}\cdot\boldsymbol{\sigma})\,\mathrm{d}\sigma =]\!]\boldsymbol{\kappa}\cdot\boldsymbol{\sigma}[\![, \\ \displaystyle\int_{\mathcal{C}} \partial_\sigma(\mathbf{A}[\boldsymbol{\sigma}])\,\mathrm{d}\sigma =]\!]\mathbf{A}[\boldsymbol{\sigma}][\![, \end{cases} \tag{81}$$

where $]\!]\boldsymbol{\kappa}\cdot\boldsymbol{\sigma}[\![:= \boldsymbol{\kappa}\cdot\boldsymbol{\sigma}|_{\partial\mathcal{C}|_{\sigma=1}} - \boldsymbol{\kappa}\cdot\boldsymbol{\sigma}|_{\partial\mathcal{C}|_{\sigma=0}}$ and $]\!]\mathbf{A}[\boldsymbol{\sigma}][\![:= \mathbf{A}[\boldsymbol{\sigma}]|_{\partial\mathcal{C}|_{\sigma=1}} - \mathbf{A}[\boldsymbol{\sigma}]|_{\partial\mathcal{C}|_{\sigma=0}}$. Note that at $\sigma = 0$, $\boldsymbol{\sigma}$ represents the inward tangent.

Owing to the lack of smoothness at a junction-point $\mathcal{O}$, for any smooth vector field $\boldsymbol{\kappa}$ and tensor field $\mathbf{A}$ on a nonsmooth closed oriented curve $\mathcal{C}$ with limiting outward unit tangent ${}^{-}\boldsymbol{\sigma}$ and ${}^{+}\boldsymbol{\sigma}$ at $\mathcal{O}$, the gradient theorem on a closed curve $\mathcal{C}$ also exhibits a *surplus*, that is,

$$\begin{cases} \displaystyle\int_{\mathcal{C}} \partial_\sigma(\boldsymbol{\kappa}\cdot\boldsymbol{\sigma})\,\mathrm{d}\sigma = [\![\boldsymbol{\kappa}\cdot\boldsymbol{\sigma}]\!], \\ \displaystyle\int_{\mathcal{C}} \partial_\sigma(\mathbf{A}[\boldsymbol{\sigma}])\,\mathrm{d}\sigma = [\![\mathbf{A}[\boldsymbol{\sigma}]]\!], \end{cases} \tag{82}$$

where $[\![\boldsymbol{\kappa}\cdot\boldsymbol{\sigma}]\!] := (\boldsymbol{\kappa}\cdot{}^{+}\boldsymbol{\sigma})|_{\mathcal{O}} - (\boldsymbol{\kappa}\cdot{}^{-}\boldsymbol{\sigma})|_{\mathcal{O}}$ and $[\![\mathbf{A}[\boldsymbol{\sigma}]]\!] := \mathbf{A}[{}^{+}\boldsymbol{\sigma}]|_{\mathcal{O}} - \mathbf{A}[{}^{-}\boldsymbol{\sigma}]|_{\mathcal{O}}$. Finally, on a nonsmooth open curve, we have that

$$\begin{cases} \displaystyle\int_{\mathcal{C}} \partial_\sigma(\boldsymbol{\kappa}\cdot\boldsymbol{\sigma})\,\mathrm{d}\sigma =]\!]\boldsymbol{\kappa}\cdot\boldsymbol{\sigma}[\![+ [\![\boldsymbol{\kappa}\cdot\boldsymbol{\sigma}]\!], \\ \displaystyle\int_{\mathcal{C}} \partial_\sigma(\mathbf{A}[\boldsymbol{\sigma}])\,\mathrm{d}\sigma =]\!]\mathbf{A}[\boldsymbol{\sigma}][\![+ [\![\mathbf{A}[\boldsymbol{\sigma}]]\!]. \end{cases} \tag{83}$$

6 Transport Relations and Mass Balance

Let a reference region $\mathcal{A}$ undergo deformation such that $\boldsymbol{y}(\mathcal{A}) := \mathcal{A}_\tau$ represents the deformed configuration, with $\boldsymbol{y}(\boldsymbol{x}, t) \in \mathcal{A}_\tau$ and $\boldsymbol{x} \in \mathcal{A}$. Through what follows, we distinguish three different regions. These regions are a material part $\mathcal{A}$, a spatial part $\mathcal{A}_\tau$, and a control volume $\overline{\mathcal{A}}$, and this classification applies for $\mathcal{A}$ being a volume, a surface, and a curve. Note that, in contrast to $\overline{\mathcal{A}}$, material cannot migrate across $\partial\mathcal{A}_\tau$.

Next, consider a spatial part $\mathcal{P}_\tau$ advecting with a smooth velocity $\boldsymbol{\kappa}$ along with the body $\mathcal{B}_\tau$. Let $\mathcal{P}$ be a material part that lies in the body $\mathcal{B}$. Let us now consider the following Jacobi's identity for an invertible second-order tensor $\mathbf{A}$, Gurtin [4],

$$\dot{\overline{|\mathbf{A}|}} = |\mathbf{A}|\,\mathrm{tr}\,(\dot{\mathbf{A}}\mathbf{A}^{-1}). \tag{84}$$

With $\mathbf{F} := \mathrm{grad}\,\boldsymbol{y}$, where $\boldsymbol{y}$ represents the motion of $\mathcal{P}_\tau$, $\mathbf{C} := \mathbf{F}^{\top}\mathbf{F}$, $\mathbf{L} := \mathrm{grad}\,\boldsymbol{\kappa}$, and $\dot{\mathbf{F}} = \mathbf{L}\mathbf{F}$, where the dot operator is the total time derivative, we have that

$$\begin{aligned}\dot{\mathbf{C}} &= \dot{\mathbf{F}}^{\top}\mathbf{F} + \mathbf{F}^{\top}\dot{\mathbf{F}},\\ &= 2\mathbf{F}^{\top}\mathbf{L}\mathbf{F}.\end{aligned} \tag{85}$$

Then, in view of (84) and (85) with $\mathrm{tr}\,(\mathbf{XYZ}) = \mathrm{tr}\,(\mathbf{YZX}) = \mathrm{tr}\,(\mathbf{ZXY})$ for any tensors $\mathbf{X}$, $\mathbf{Y}$, and $\mathbf{Z}$, we obtain

$$\begin{aligned}\dot{\overline{|\mathbf{C}|}} &= |\mathbf{C}|\,\mathrm{tr}\,(2\mathbf{F}^{\top}\mathbf{L}\mathbf{F}\mathbf{F}^{-1}\mathbf{F}^{-\top}),\\ &= |\mathbf{C}|\,\mathrm{tr}\,(2\mathbf{F}^{\top}\mathbf{L}\mathbf{F}^{-\top}),\\ &= |\mathbf{C}|\,\mathrm{tr}\,(2\mathbf{F}^{-\top}\mathbf{F}^{\top}\mathbf{L}),\\ &= |\mathbf{C}|\,\mathrm{tr}\,(2\mathbf{L}),\\ &= 2|\mathbf{C}|\,\mathrm{div}\,\boldsymbol{\kappa}.\end{aligned} \tag{86}$$

Finally, we have the volumetric Jacobian of deformation defined as

$$J := \frac{\mathrm{d}v_\tau}{\mathrm{d}v} = \sqrt{|\mathbf{C}|}. \tag{87}$$

Also, bear in mind that $\dot{\overline{J^2}} = 2J\dot{J}$, which implies

$$\dot{J} = J\,\mathrm{div}\,\boldsymbol{\kappa}. \tag{88}$$

Then, given a smooth scalar field ϕ on $\mathcal{P}_\tau$, we have that

$$\overline{\int_{\mathcal{P}_\tau} \phi \, \mathrm{d}v_\tau}^{\boldsymbol{\cdot}} = \int_{\mathcal{P}} (\dot{\phi} J + \phi \dot{J}) \, \mathrm{d}v,$$
$$= \int_{\mathcal{P}} (\dot{\phi} + \phi \operatorname{div} \boldsymbol{\kappa}) J \, \mathrm{d}v,$$
$$= \int_{\mathcal{P}_\tau} (\dot{\phi} + \phi \operatorname{div} \boldsymbol{\kappa}) \, \mathrm{d}v_\tau. \tag{89}$$

We now consider that the total mass of a system reads

$$\int_{\mathcal{P}_\tau} \varrho(\boldsymbol{y}_\tau, t) \, \mathrm{d}v_\tau = \int_{\mathcal{P}} \varrho_0(\boldsymbol{x}) \, \mathrm{d}v = \text{constant}, \tag{90}$$

where ϱ and ϱ_0 are, respectively, the densities in the spatial and reference frames. Noting that the right-hand-side of (90) does not depend on time, requiring the total mass to be balanced is equivalent to

$$\overline{\int_{\mathcal{P}_\tau} \varrho(\boldsymbol{y}_\tau, t) \, \mathrm{d}v_\tau}^{\boldsymbol{\cdot}} = 0, \tag{91}$$

and using (89), we arrive at the partwise mass balance

$$\int_{\mathcal{P}_\tau} (\dot{\varrho} + \varrho \operatorname{div} \boldsymbol{\kappa}) \, \mathrm{d}v_\tau = 0, \tag{92}$$

which renders the following pointwise mass balance

$$\dot{\varrho} + \varrho \operatorname{div} \boldsymbol{\kappa} = 0. \tag{93}$$

An immediate consequence of Reynolds' transport theorem with the balance of mass is that

$$\overline{\int_{\mathcal{P}_\tau} \varrho\phi \, \mathrm{d}v_\tau}^{\boldsymbol{\cdot}} = \int_{\mathcal{P}_\tau} (\dot{\varrho}\phi + \varrho\dot{\phi} + \varrho\phi \operatorname{div} \boldsymbol{\kappa}) \, \mathrm{d}v_\tau,$$
$$= \int_{\mathcal{P}_\tau} \varrho\dot{\phi} \, \mathrm{d}v_\tau. \tag{94}$$

Next, owing to the mass balance, consider the following identities for a scalar field ϕ and a vector field $\boldsymbol{\phi}$, or a tensor of order greater than one,

$$\varrho\dot{\phi} = \partial_t(\varrho\phi) + \operatorname{div}(\varrho\phi\,\boldsymbol{\kappa}), \tag{95}$$

and

$$\varrho \dot{\boldsymbol{\phi}} = \partial_t(\varrho \boldsymbol{\phi}) + \operatorname{div}(\varrho \boldsymbol{\phi} \otimes \boldsymbol{\kappa}). \tag{96}$$

Integrating on a control volume expressions (95) and (96) and applying the divergence theorem, we have that

$$\begin{cases} \displaystyle\int_{\overline{\mathcal{P}}} \varrho \dot{\phi} \, \mathrm{d}v = \overline{\int_{\overline{\mathcal{P}}} \varrho \phi \, \mathrm{d}v}^{\boldsymbol{\cdot}} + \int_{\partial \overline{\mathcal{P}}} \varrho \phi \, \boldsymbol{\kappa} \cdot \boldsymbol{n} \, \mathrm{d}a, \\ \displaystyle\int_{\overline{\mathcal{P}}} \varrho \dot{\boldsymbol{\phi}} \, \mathrm{d}v = \overline{\int_{\overline{\mathcal{P}}} \varrho \boldsymbol{\phi} \, \mathrm{d}v}^{\boldsymbol{\cdot}} + \int_{\partial \overline{\mathcal{P}}} \varrho \, (\boldsymbol{\phi} \otimes \boldsymbol{\kappa})[\boldsymbol{n}] \, \mathrm{d}a, \end{cases} \tag{97}$$

where we have used that

$$\int_{\overline{\mathcal{P}}} \partial_t(\varrho \phi) \, \mathrm{d}v = \overline{\int_{\overline{\mathcal{P}}} \varrho \phi \, \mathrm{d}v}^{\boldsymbol{\cdot}} \qquad \text{and} \qquad \int_{\overline{\mathcal{P}}} \partial_t(\varrho \boldsymbol{\phi}) \, \mathrm{d}v = \overline{\int_{\overline{\mathcal{P}}} \varrho \boldsymbol{\phi} \, \mathrm{d}v}^{\boldsymbol{\cdot}}, \tag{98}$$

since $\overline{\mathcal{P}}$ does not depend on t.

Lastly, by setting $\phi = 1$ in (97), we arrive at the mass balance for a control volume,

$$\overline{\int_{\overline{\mathcal{P}}} \varrho \, \mathrm{d}v}^{\boldsymbol{\cdot}} = -\int_{\partial \overline{\mathcal{P}}} \varrho \, \boldsymbol{\kappa} \cdot \boldsymbol{n} \, \mathrm{d}a. \tag{99}$$

Part II
The Principle of Virtual Powers and Complementary Balances

Chapter 3
Power Balance, Fields, and Hyperfields

The principle of virtual powers is a powerful tool that can be traced back prior to Johann Bernoulli, whose formalism of the principle is well known.[1] Nowadays, it is probably the most prevalent tool when it comes to deriving mechanical theories for enriched continua. However, it was not until the two works by Toupin [15, 16] that *nonsimple materials* were in focus. According to Edelen [11], Voigt [9] thought and elaborated the idea of nonlocalities to augment the hypothesis of Euler–Cauchy. Analogously, using the principle of virtual power version postulated by Gurtin [42], Fried & Gurtin [20] generalized Toupin's work to account for general second-grade materials while Podio-Guidugli [43] worked on third-grade hyperelastic solids. We base our treatment on what is postulated next, the principle of virtual powers. In this chapter, we study the consequences of this principle, namely, the pointwise field equations in conjunction with the microtractions, tractions, hypermicrotractions, hypertractions, microforces, and forces.

1 The Principle of Virtual Powers on $\overline{\mathcal{P}} \subset \mathcal{B}$

In this work, the principle of virtual powers for an arbitrary control volume $\overline{\mathcal{P}}$ is postulated as follows.

Postulate 1 (Principle of virtual powers) *Considering motion accompanied by n transition layers, the principle of virtual power states that the following partwise balances of virtual powers*

[1] Capecchi [50] mentions that, to some extent, the principle was studied by Latin medieval scholars, ancient Greeks, and ancient Arabs. For a broad overview of the principle, the reader is also referred to the works by Lidström [51], which includes the treatment of shocks as well as external and internal constraints, and by Del Piero [52], including first and second gradient continuum theories.

L. Espath, *Mechanics and Geometry of Enriched Continua*,
https://doi.org/10.1007/978-3-031-28934-7_3

$$\mathcal{V}^{\flat}_{\text{int}}(\overline{\mathcal{P}}; \{\chi^{\alpha}\}^{n}_{\alpha=1}) = \mathcal{V}^{\flat}_{\text{ext}}(\overline{\mathcal{P}}; \{\chi^{\alpha}\}^{n}_{\alpha=1}) \tag{100}$$

and

$$\mathcal{V}^{\sharp}_{\text{int}}(\overline{\mathcal{P}}; \boldsymbol{\chi}) = \mathcal{V}^{\sharp}_{\text{ext}}(\overline{\mathcal{P}}; \boldsymbol{\chi}) \tag{101}$$

hold for any arbitrary control volume $\overline{\mathcal{P}} \subset \mathcal{B}$, time t, and any choice of smooth admissible scalar χ^{α} and vector $\boldsymbol{\chi}$ virtual fields. For the scalar virtual fields, the internal virtual power $\mathcal{V}^{\flat}_{\text{int}}$ is

$$\mathcal{V}^{\flat}_{\text{int}}(\overline{\mathcal{P}}; \{\chi^{\alpha}\}^{n}_{\alpha=1}) := \sum_{\alpha=1}^{n} \bigg\{ \int\limits_{\overline{\mathcal{P}}} \big(-\pi^{\alpha}\chi^{\alpha} + \boldsymbol{\xi}^{\alpha} \cdot \operatorname{grad}\chi^{\alpha} + \boldsymbol{\Xi}^{\alpha} : \operatorname{grad}^{2}\chi^{\alpha} + \mathbb{X}^{\alpha} \,\vdots\, \operatorname{grad}^{3}\chi^{\alpha} \big)\, \mathrm{d}v \bigg\}, \tag{102}$$

and the external virtual power $\mathcal{V}^{\flat}_{\text{ext}}$ is

$$\begin{aligned}\mathcal{V}^{\flat}_{\text{ext}}(\overline{\mathcal{P}}; \{\chi^{\alpha}\}^{n}_{\alpha=1}) := \sum_{\alpha=1}^{n} \bigg\{ &\int\limits_{\overline{\mathcal{P}}} \gamma^{\alpha}\chi^{\alpha}\, \mathrm{d}v + \int\limits_{\partial\overline{\mathcal{P}}} \big(\xi^{\alpha}_{\mathcal{S}}\chi^{\alpha} + {}^{2}\xi^{\alpha}_{\mathcal{S}}\partial_{n}\chi^{\alpha} + {}^{3}\xi^{\alpha}_{\mathcal{S}}\partial^{2}_{n}\chi^{\alpha}\big)\, \mathrm{d}a \\ &+ \int\limits_{\overline{\mathcal{C}}} \big(\xi^{\alpha}_{\mathcal{C}}\chi^{\alpha} + {}^{2}_{n}\xi^{\alpha}_{\mathcal{C}}\partial_{n}\chi^{\alpha} + {}^{2}_{\nu}\xi^{\alpha}_{\mathcal{C}}\partial_{\nu}\chi^{\alpha}\big)\, \mathrm{d}\sigma + \xi^{\alpha}_{\mathcal{O}}\chi^{\alpha}|_{\mathcal{O}} \bigg\},\end{aligned} \tag{103}$$

whereas for the vector virtual field, the internal virtual power $\mathcal{V}^{\sharp}_{\text{int}}$ is

$$\mathcal{V}^{\sharp}_{\text{int}}(\overline{\mathcal{P}}; \boldsymbol{\chi}) := \int\limits_{\overline{\mathcal{P}}} \big(\mathbf{T} : \operatorname{grad}\boldsymbol{\chi} + \mathbb{T} \,\vdots\, \operatorname{grad}^{2}\boldsymbol{\chi} + \mathbf{T} :: \operatorname{grad}^{3}\boldsymbol{\chi}\big)\, \mathrm{d}v, \tag{104}$$

and the external virtual power $\mathcal{V}^{\sharp}_{\text{ext}}$ is

$$\begin{aligned}\mathcal{V}^{\sharp}_{\text{ext}}(\overline{\mathcal{P}}; \boldsymbol{\chi}) := &\int\limits_{\overline{\mathcal{P}}} \boldsymbol{b} \cdot \boldsymbol{\chi}\, \mathrm{d}v + \int\limits_{\partial\overline{\mathcal{P}}} \big(\boldsymbol{t}_{\mathcal{S}} \cdot \boldsymbol{\chi} + {}^{2}\boldsymbol{t}_{\mathcal{S}} \cdot \partial_{n}\boldsymbol{\chi} + {}^{3}\boldsymbol{t}_{\mathcal{S}} \cdot \partial^{2}_{n}\boldsymbol{\chi}\big)\, \mathrm{d}a \\ &+ \int\limits_{\overline{\mathcal{C}}} \big(\boldsymbol{t}_{\mathcal{C}} \cdot \boldsymbol{\chi} + {}^{2}_{n}\boldsymbol{t}_{\mathcal{C}} \cdot \partial_{n}\boldsymbol{\chi} + {}^{2}_{\nu}\boldsymbol{t}_{\mathcal{C}} \cdot \partial_{\nu}\boldsymbol{\chi}\big)\, \mathrm{d}\sigma + \boldsymbol{t}_{\mathcal{O}} \cdot \boldsymbol{\chi}|_{\mathcal{O}}.\end{aligned} \tag{105}$$

Here, the αth surface microtraction $\xi^{\alpha}_{\mathcal{S}}$, the αth surface hypertraction ${}^{2}\xi^{\alpha}_{\mathcal{S}}$, and the αth surface supra-hypermicrotraction ${}^{3}\xi^{\alpha}_{\mathcal{S}}$ expend power per unit area conjugate to χ^{α}, $\partial_{n}\chi^{\alpha}$, and $\partial^{2}_{n}\chi^{\alpha}$, respectively. The αth junction-edge microtraction $\xi^{\alpha}_{\mathcal{C}}$ and the

αth junction-edge hypermicrotractions ${}^{2}_{n}\xi^{\alpha}_{\mathcal{C}}$ and ${}^{2}_{\nu}\xi^{\alpha}_{\mathcal{C}}$ expend power per unit length conjugate to χ^{α}, $\partial_n \chi^{\alpha}$, and $\partial_\nu \chi^{\alpha}$, respectively. Lastly, the αth junction-point microforce $\xi^{\alpha}_{\mathcal{O}}$ expends power conjugate to χ^{α}. Analogously, the surface traction $\boldsymbol{t}_{\mathcal{S}}$, the surface hypertraction ${}^{2}\boldsymbol{t}_{\mathcal{S}}$, and the surface supra-hypertraction ${}^{3}\boldsymbol{t}_{\mathcal{S}}$ expend power per unit area conjugate to $\boldsymbol{\chi}$, $\partial_n \boldsymbol{\chi}$, and $\partial^2_n \boldsymbol{\chi}$, respectively. The junction-edge traction $\boldsymbol{t}_{\mathcal{C}}$ and the junction-edge hypertractions ${}^{2}_{n}\boldsymbol{t}_{\mathcal{C}}$ and ${}^{2}_{\nu}\boldsymbol{t}_{\mathcal{C}}$ expend power per unit length conjugate to $\boldsymbol{\chi}$, $\partial_n \boldsymbol{\chi}$, and $\partial_\nu \boldsymbol{\chi}$, respectively. Lastly, the junction-point force $\boldsymbol{t}_{\mathcal{O}}$ expends power conjugate to $\boldsymbol{\chi}$.

Next, we aim to derive the explicit form of each microtraction, microforce, traction, and force involved in these balances of powers. As a byproduct, we also obtain the field equations resulting from the principle of virtual powers. Now, consider the following identities, respectively, related to the last three terms of internal power expenditure (102),

$$\boldsymbol{\xi}^{\alpha} \cdot \operatorname{grad} \chi^{\alpha} = \operatorname{div}(\boldsymbol{\xi}^{\alpha} \chi^{\alpha}) - \chi^{\alpha} \operatorname{div} \boldsymbol{\xi}^{\alpha}, \tag{106}$$

$$\begin{aligned} \boldsymbol{\Xi}^{\alpha} : \operatorname{grad}^2 \chi^{\alpha} &= \operatorname{div}((\boldsymbol{\Xi}^{\alpha})^{\top}[\operatorname{grad} \chi^{\alpha}]) - (\operatorname{div} \boldsymbol{\Xi}^{\alpha}) \cdot \operatorname{grad} \chi^{\alpha}, \\ &= \operatorname{div}((\boldsymbol{\Xi}^{\alpha})^{\top}[\operatorname{grad} \chi^{\alpha}]) - \operatorname{div}((\operatorname{div} \boldsymbol{\Xi}^{\alpha}) \chi^{\alpha}) + \chi^{\alpha} \operatorname{div}^2 \boldsymbol{\Xi}^{\alpha}, \end{aligned} \tag{107}$$

and

$$\begin{aligned} \mathbb{X}^{\alpha} \vdots \operatorname{grad}^3 \chi^{\alpha} &= \operatorname{div}((\mathbb{X}^{\alpha})^{\dashv}[\operatorname{grad}^2 \chi^{\alpha}]) - (\operatorname{div} \mathbb{X}^{\alpha}) : \operatorname{grad}^2 \chi^{\alpha}, \\ &= \operatorname{div}((\mathbb{X}^{\alpha})^{\dashv}[\operatorname{grad}^2 \chi^{\alpha}]) - \operatorname{div}((\operatorname{div} \mathbb{X}^{\alpha})^{\top}[\operatorname{grad} \chi^{\alpha}]) \\ &\quad + (\operatorname{div}^2 \mathbb{X}^{\alpha}) \cdot \operatorname{grad} \chi^{\alpha}, \\ &= \operatorname{div}((\mathbb{X}^{\alpha})^{\dashv}[\operatorname{grad}^2 \chi^{\alpha}]) - \operatorname{div}((\operatorname{div} \mathbb{X}^{\alpha})^{\top}[\operatorname{grad} \chi^{\alpha}]) \\ &\quad + \operatorname{div}((\operatorname{div}^2 \mathbb{X}^{\alpha}) \chi^{\alpha}) - \chi^{\alpha} \operatorname{div}^3 \mathbb{X}^{\alpha}. \end{aligned} \tag{108}$$

Also, for second- and third-order smooth tensor fields $\mathbf{A}$ and $\mathbb{A}$, respectively, with symmetries

$$\begin{cases} \mathbf{A}[\boldsymbol{a}_1] \cdot \boldsymbol{a}_2 & \text{unaltered under the permutation of } \boldsymbol{a}_1 \text{ and } \boldsymbol{a}_2, \\ \mathbb{A}[\boldsymbol{a}_1][\boldsymbol{a}_2] \cdot \boldsymbol{a}_3 & \text{unaltered under any permutation of } \boldsymbol{a}_1, \boldsymbol{a}_2 \text{ and } \boldsymbol{a}_3, \end{cases} \tag{109}$$

consider the identities associated with $\mathbf{A}[\boldsymbol{n}] \cdot \operatorname{grad} \kappa$ and $\mathbb{A}[\boldsymbol{n}] : \operatorname{grad}^2 \kappa$ as follows.

$$\begin{aligned} \mathbf{A}[\boldsymbol{n}] \cdot \operatorname{grad} \kappa &= \boldsymbol{n} \cdot \mathbf{A}[\boldsymbol{n}] \partial_n \kappa + \mathbf{A}[\boldsymbol{n}] \cdot \operatorname{grad}_{\mathcal{S}} \kappa, \\ &= \boldsymbol{n} \cdot \mathbf{A}[\boldsymbol{n}] \partial_n \kappa + \operatorname{div}_{\mathcal{S}}(\mathbf{P}_{\!n}[\mathbf{A}[\boldsymbol{n}]]\kappa) \\ &\quad - \operatorname{div}_{\mathcal{S}}(\mathbf{A}[\boldsymbol{n}])\kappa - 2K\boldsymbol{n} \cdot \mathbf{A}[\boldsymbol{n}]\kappa, \end{aligned} \tag{110}$$

and

$$
\begin{aligned}
\mathbb{A}[\boldsymbol{n}] : \operatorname{grad}^2\kappa &= \mathbb{A}[\boldsymbol{n}] : \operatorname{grad}_{\mathcal{S}}^2\kappa + \mathbb{A}[\boldsymbol{n}][\boldsymbol{n}] \cdot \mathbf{K}^{\top}[\operatorname{grad}_{\mathcal{S}}\kappa] \\
&\quad + 2\,\mathbb{A}[\boldsymbol{n}][\boldsymbol{n}] \cdot \operatorname{grad}_{\mathcal{S}}(\partial_n\kappa) - \partial_n\kappa\,\mathbb{A}[\boldsymbol{n}] : \mathbf{K} \\
&\quad + \partial_n^2\kappa\,\boldsymbol{n} \cdot \mathbb{A}[\boldsymbol{n}][\boldsymbol{n}], \\
&= \operatorname{div}_{\mathcal{S}}((\mathbb{A}[\boldsymbol{n}]\mathbf{P}_{\boldsymbol{n}})^{\top}[\operatorname{grad}_{\mathcal{S}}\kappa]) \\
&\quad - \operatorname{div}_{\mathcal{S}}(\mathbf{P}_{\boldsymbol{n}}[\operatorname{div}_{\mathcal{S}}(\mathbb{A}[\boldsymbol{n}]) + 2K\,\mathbb{A}[\boldsymbol{n}][\boldsymbol{n}]]\kappa) \\
&\quad + \operatorname{div}_{\mathcal{S}}(\operatorname{div}_{\mathcal{S}}(\mathbb{A}[\boldsymbol{n}]) + 2K\,\mathbb{A}[\boldsymbol{n}][\boldsymbol{n}])\kappa \\
&\quad + 2K\boldsymbol{n} \cdot (\operatorname{div}_{\mathcal{S}}(\mathbb{A}[\boldsymbol{n}]) + 2K\,\mathbb{A}[\boldsymbol{n}][\boldsymbol{n}])\kappa \\
&\quad + \operatorname{div}_{\mathcal{S}}(\mathbb{A}[\boldsymbol{n}][\boldsymbol{n}] \cdot \mathbf{K}^{\top}\kappa) - \operatorname{div}_{\mathcal{S}}(\mathbb{A}[\boldsymbol{n}][\boldsymbol{n}] \cdot \mathbf{K}^{\top})\kappa \\
&\quad + 2\operatorname{div}_{\mathcal{S}}(\mathbf{P}_{\boldsymbol{n}}[\mathbb{A}[\boldsymbol{n}][\boldsymbol{n}]]\partial_n\kappa) - 2\operatorname{div}_{\mathcal{S}}(\mathbb{A}[\boldsymbol{n}][\boldsymbol{n}])\partial_n\kappa \\
&\quad - 4K\boldsymbol{n} \cdot \mathbb{A}[\boldsymbol{n}][\boldsymbol{n}]\partial_n\kappa - \partial_n\kappa\,\mathbb{A}[\boldsymbol{n}] : \mathbf{K} + \partial_n^2\kappa\,\boldsymbol{n} \cdot \mathbb{A}[\boldsymbol{n}][\boldsymbol{n}]. \qquad (111)
\end{aligned}
$$

Note that the knowledge of $\boldsymbol{n}^+$ and $\boldsymbol{n}^-$ of a junction-edge $\mathcal{C}$ implies the knowledge of $\boldsymbol{\nu}^+$ and $\boldsymbol{\nu}^-$, since they all live in the same plane. This fact is relevant when it comes to defining the arguments of each traction.

Now, replacing identities (106)–(108) into (102) and then applying the divergence theorem, we are in a position to use identities (110) with $\mathbf{A}[\boldsymbol{n}] := (\boldsymbol{\Xi}^{\alpha} - \operatorname{div}\mathbb{X}^{\alpha})[\boldsymbol{n}]$ and (111) with $\mathbb{A}[\boldsymbol{n}] := \mathbb{X}^{\alpha}[\boldsymbol{n}]$. Next, using the surface divergence theorem (75) for nonsmooth closed surfaces followed by the gradient theorem $(82)_1$ for nonsmooth closed curves, and summing over α, we arrive at

$$
\begin{aligned}
&\sum_{\alpha=1}^{n} \int_{\mathcal{P}} \big(-\pi^{\alpha}\chi^{\alpha} + \boldsymbol{\xi}^{\alpha} \cdot \operatorname{grad}\chi^{\alpha} + \boldsymbol{\Xi}^{\alpha} : \operatorname{grad}^2\chi^{\alpha} + \mathbb{X}^{\alpha} \,\vdots\, \operatorname{grad}^3\chi^{\alpha} \big)\, \mathrm{d}v \\
&= \sum_{\alpha=1}^{n} \bigg\{ \int_{\mathcal{P}} \chi^{\alpha} \big(-\pi^{\alpha} - \operatorname{div}\boldsymbol{\xi}^{\alpha} + \operatorname{div}^2\boldsymbol{\Xi}^{\alpha} - \operatorname{div}^3\mathbb{X}^{\alpha} \big)\, \mathrm{d}v \\
&\quad + \int_{\partial\mathcal{P}} \big(\bar{\xi}_{\mathcal{S}}^{\alpha}\chi^{\alpha} + {}^{2}\bar{\xi}_{\mathcal{S}}^{\alpha}\partial_n\chi^{\alpha} + {}^{3}\bar{\xi}_{\mathcal{S}}^{\alpha}\partial_n^2\chi^{\alpha} \big)\, \mathrm{d}a \\
&\quad + \int_{\mathcal{C}} \big(\bar{\xi}_{\mathcal{C}}^{\alpha}\chi^{\alpha} + {}_{n}^{2}\bar{\xi}_{\mathcal{C}}^{\alpha}\partial_n\chi^{\alpha} + {}_{\nu}^{2}\bar{\xi}_{\mathcal{C}}^{\alpha}\partial_{\nu}\chi^{\alpha} \big)\, \mathrm{d}\sigma + \bar{\xi}_{\mathcal{O}}^{\alpha}\chi^{\alpha}|_{\mathcal{O}} \bigg\}, \qquad (112)
\end{aligned}
$$

where $\bar{\xi}_{\mathcal{S}}^{\alpha}$, ${}^{2}\bar{\xi}_{\mathcal{S}}^{\alpha}$, ${}^{3}\bar{\xi}_{\mathcal{S}}^{\alpha}$, $\bar{\xi}_{\mathcal{C}}^{\alpha}$, ${}_{n}^{2}\bar{\xi}_{\mathcal{C}}^{\alpha}$, ${}_{\nu}^{2}\bar{\xi}_{\mathcal{C}}^{\alpha}$, and $\bar{\xi}_{\mathcal{O}}^{\alpha}$ are given as follows.

$$
\begin{aligned}
\bar{\xi}^{\alpha}_{\mathcal{S}}(\boldsymbol{x}, t; \boldsymbol{n}, \mathbf{K}, \mathbb{K}) := (\boldsymbol{\xi}^{\alpha} - \operatorname{div} \boldsymbol{\Xi}^{\alpha} + \operatorname{div}^2 \mathbb{X}^{\alpha}) \cdot \boldsymbol{n} \\
- \operatorname{div}_{\mathcal{S}}((\boldsymbol{\Xi}^{\alpha} - \operatorname{div} \mathbb{X}^{\alpha})[\boldsymbol{n}]) - 2K\, \boldsymbol{n} \cdot (\boldsymbol{\Xi}^{\alpha} - \operatorname{div} \mathbb{X}^{\alpha})[\boldsymbol{n}] \\
+ \operatorname{div}_{\mathcal{S}}(\operatorname{div}_{\mathcal{S}}(\mathbb{X}^{\alpha}[\boldsymbol{n}]) + 2K\, \mathbb{X}^{\alpha}[\boldsymbol{n}][\boldsymbol{n}]) \\
+ 2K\, \boldsymbol{n} \cdot (\operatorname{div}_{\mathcal{S}}(\mathbb{X}^{\alpha}[\boldsymbol{n}]) + 2K\, \mathbb{X}^{\alpha}[\boldsymbol{n}][\boldsymbol{n}]) \\
- \operatorname{div}_{\mathcal{S}}(\mathbb{X}^{\alpha}[\boldsymbol{n}][\boldsymbol{n}] \cdot \mathbf{K}^{\top}),
\end{aligned}
\tag{113}
$$

$$
\begin{aligned}
{}^{2}\bar{\xi}^{\alpha}_{\mathcal{S}}(\boldsymbol{x}, t; \boldsymbol{n}, \mathbf{K}) := \boldsymbol{n} \cdot (\boldsymbol{\Xi}^{\alpha} - \operatorname{div} \mathbb{X}^{\alpha})[\boldsymbol{n}] - 2 \operatorname{div}_{\mathcal{S}}(\mathbb{X}^{\alpha}[\boldsymbol{n}][\boldsymbol{n}]) \\
- 4K\, \boldsymbol{n} \cdot \mathbb{X}^{\alpha}[\boldsymbol{n}][\boldsymbol{n}] - \mathbb{X}^{\alpha}[\boldsymbol{n}] : \mathbf{K},
\end{aligned}
\tag{114}
$$

$$
{}^{3}\bar{\xi}^{\alpha}_{\mathcal{S}}(\boldsymbol{x}, t; \boldsymbol{n}) := \boldsymbol{n} \cdot \mathbb{X}^{\alpha}[\boldsymbol{n}][\boldsymbol{n}], \tag{115}
$$

$$
\begin{aligned}
\bar{\xi}^{\alpha}_{\mathcal{C}}(\boldsymbol{x}, t; \boldsymbol{n}^{\pm}, \mathbf{K}^{\pm}) := \{\!\!\{ \boldsymbol{\nu} \cdot ((\boldsymbol{\Xi}^{\alpha} - \operatorname{div} \mathbb{X}^{\alpha})[\boldsymbol{n}] - \operatorname{div}_{\mathcal{S}}(\mathbb{X}^{\alpha}[\boldsymbol{n}]) \\
- 2K\, \mathbb{X}^{\alpha}[\boldsymbol{n}][\boldsymbol{n}] + \mathbb{X}^{\alpha}[\boldsymbol{n}][\boldsymbol{n}] \cdot \mathbf{K}^{\top}) \\
- \partial_{\sigma}(\boldsymbol{\sigma} \cdot \mathbb{X}^{\alpha}[\boldsymbol{n}][\boldsymbol{\nu}]) \}\!\!\},
\end{aligned}
\tag{116}
$$

$$
{}^{2}_{n}\bar{\xi}^{\alpha}_{\mathcal{C}}(\boldsymbol{x}, t; \boldsymbol{n}^{\pm}) := \{\!\!\{ 2\, \boldsymbol{\nu} \cdot \mathbb{X}^{\alpha}[\boldsymbol{n}][\boldsymbol{n}] \}\!\!\}, \tag{117}
$$

$$
{}^{2}_{\nu}\bar{\xi}^{\alpha}_{\mathcal{C}}(\boldsymbol{x}, t; \boldsymbol{n}^{\pm}) := \{\!\!\{ \boldsymbol{\nu} \cdot \mathbb{X}^{\alpha}[\boldsymbol{n}][\boldsymbol{\nu}] \}\!\!\}, \tag{118}
$$

and

$$
\bar{\xi}^{\alpha}_{\mathcal{O}}(\boldsymbol{x}, t; {}^{\pm}\boldsymbol{n}^{\pm}) := [\![\{\!\!\{ \boldsymbol{\sigma} \cdot \mathbb{X}^{\alpha}[\boldsymbol{n}][\boldsymbol{\nu}] \}\!\!\}]\!]. \tag{119}
$$

Thus, we have that the principle of virtual powers (100) is satisfied for any arbitrary control volume $\overline{\mathcal{P}} \subset \mathcal{B}$, time t, and any choice of smooth admissible scalars χ^{α}, if and only if

$$
\begin{aligned}
\sum_{\alpha=1}^{n} \Bigg\{ \int_{\overline{\mathcal{P}}} \chi^{\alpha} \big(\gamma^{\alpha} + \pi^{\alpha} + \operatorname{div} \boldsymbol{\xi}^{\alpha} - \operatorname{div}^2 \boldsymbol{\Xi}^{\alpha} + \operatorname{div}^3 \mathbb{X}^{\alpha}\big)\, \mathrm{d}v \\
+ \int_{\partial \overline{\mathcal{P}}} \big((\xi^{\alpha}_{\mathcal{S}} - \bar{\xi}^{\alpha}_{\mathcal{S}})\chi^{\alpha} + ({}^{2}\xi^{\alpha}_{\mathcal{S}} - {}^{2}\bar{\xi}^{\alpha}_{\mathcal{S}})\partial_n \chi^{\alpha} + ({}^{3}\xi^{\alpha}_{\mathcal{S}} - {}^{3}\bar{\xi}^{\alpha}_{\mathcal{S}})\partial_n^2 \chi^{\alpha}\big)\, \mathrm{d}a \\
+ \int_{\overline{\mathcal{C}}} \big((\xi_{\mathcal{C}} - \bar{\xi}^{\alpha}_{\mathcal{C}})\chi^{\alpha} + ({}^{2}_{n}\xi^{\alpha}_{\mathcal{C}} - {}^{2}_{n}\bar{\xi}^{\alpha}_{\mathcal{C}})\partial_n \chi^{\alpha} + ({}^{2}_{\nu}\xi^{\alpha}_{\mathcal{C}} - {}^{2}_{\nu}\bar{\xi}^{\alpha}_{\mathcal{C}})\partial_{\nu} \chi^{\alpha}\big)\, \mathrm{d}\sigma \\
+ (\xi_{\mathcal{O}} - \bar{\xi}^{\alpha}_{\mathcal{O}})\chi^{\alpha}|_{\mathcal{O}} \Bigg\} = 0,
\end{aligned}
\tag{120}
$$

which results from combining (112) and (103) through (100). Moreover, given the arbitrariness and independence among the kinematic processes χ^α, $\partial_n\chi^\alpha$, $\partial_n^2\chi^\alpha$, and $\partial_\nu\chi^\alpha$, we have that $\xi_{\mathcal{S}} := \bar{\xi}^\alpha_{\mathcal{S}}$, ${}^2\xi^\alpha_{\mathcal{S}} := {}^2\bar{\xi}^\alpha_{\mathcal{S}}$, ${}^3\xi^\alpha_{\mathcal{S}} := {}^3\bar{\xi}^\alpha_{\mathcal{S}}$, $\xi_{\mathcal{C}} := \bar{\xi}^\alpha_{\mathcal{C}}$, ${}^2_n\xi^\alpha_{\mathcal{C}} := {}^2_n\bar{\xi}^\alpha_{\mathcal{C}}$, ${}^2_\nu\xi^\alpha_{\mathcal{C}} := {}^2_\nu\bar{\xi}^\alpha_{\mathcal{C}}$, and $\xi_{\mathcal{O}} := \bar{\xi}^\alpha_{\mathcal{O}}$, and that the pointwise field equation of microforces is given by

$$\operatorname{div}(\boldsymbol{\xi}^\alpha - \operatorname{div}(\boldsymbol{\Xi}^\alpha - \operatorname{div}\mathbb{X}^\alpha)) + \pi^\alpha + \gamma^\alpha = 0, \qquad \forall\, 1 \le \alpha \le n. \tag{121}$$

Reversing arguments, we may also depart from (120) to arrive at (100); then, concluding our proof.

Analogously, consider the following identities, respectively, related to the three terms of the internal power expenditure (104),

$$\mathbf{T} : \operatorname{grad}\boldsymbol{\chi} = \operatorname{div}(\mathbf{T}^{\top}[\boldsymbol{\chi}]) - (\operatorname{div}\mathbf{T}) \cdot \boldsymbol{\chi}, \tag{122}$$

$$\begin{aligned} \mathbb{T} \vdots \operatorname{grad}^2\boldsymbol{\chi} &= \operatorname{div}(\mathbb{T}^{\dashv}[\operatorname{grad}\boldsymbol{\chi}]) - (\operatorname{div}\mathbb{T}) : \operatorname{grad}\boldsymbol{\chi}, \\ &= \operatorname{div}(\mathbb{T}^{\dashv}[\operatorname{grad}\boldsymbol{\chi}]) - \operatorname{div}((\operatorname{div}\mathbb{T})^{\top}[\boldsymbol{\chi}]) + (\operatorname{div}^2\mathbb{T}) \cdot \boldsymbol{\chi}, \end{aligned} \tag{123}$$

and

$$\begin{aligned} \boldsymbol{\mathbb{T}} :: \operatorname{grad}^3\boldsymbol{\chi} &= \operatorname{div}(\boldsymbol{\mathbb{T}}^{\dashv}[\operatorname{grad}^2\boldsymbol{\chi}]) - (\operatorname{div}\boldsymbol{\mathbb{T}}) \vdots \operatorname{grad}^2\boldsymbol{\chi}, \\ &= \operatorname{div}(\boldsymbol{\mathbb{T}}^{\dashv}[\operatorname{grad}^2\boldsymbol{\chi}]) - \operatorname{div}((\operatorname{div}\boldsymbol{\mathbb{T}})^{\dashv}[\operatorname{grad}\boldsymbol{\chi}]) \\ &\quad + (\operatorname{div}^2\boldsymbol{\mathbb{T}}) : \operatorname{grad}\boldsymbol{\chi}, \\ &= \operatorname{div}(\boldsymbol{\mathbb{T}}^{\dashv}[\operatorname{grad}^2\boldsymbol{\chi}]) - \operatorname{div}((\operatorname{div}\boldsymbol{\mathbb{T}})^{\dashv}[\operatorname{grad}\boldsymbol{\chi}]) \\ &\quad + \operatorname{div}((\operatorname{div}^2\boldsymbol{\mathbb{T}})^{\top}[\boldsymbol{\chi}]) - (\operatorname{div}^3\boldsymbol{\mathbb{T}}) \cdot \boldsymbol{\chi}. \end{aligned} \tag{124}$$

Also, for third- and fourth-order smooth tensor fields $\mathbb{A}$ and $\boldsymbol{\mathbb{A}}$, respectively, with symmetries

$$\begin{cases} \mathbb{A}[\boldsymbol{a}_1][\boldsymbol{a}_2] & \text{unaltered under the permutation of } \boldsymbol{a}_1 \text{ and } \boldsymbol{a}_2, \\ \boldsymbol{\mathbb{A}}[\boldsymbol{a}_1][\boldsymbol{a}_2][\boldsymbol{a}_3] & \text{unaltered under any permutation of } \boldsymbol{a}_1, \boldsymbol{a}_2 \text{ and } \boldsymbol{a}_3, \end{cases} \tag{125}$$

consider the identities associated with $\mathbb{A}[\boldsymbol{n}] : \operatorname{grad}\boldsymbol{\kappa}$ and $\boldsymbol{\mathbb{A}}[\boldsymbol{n}] \vdots \operatorname{grad}^2\boldsymbol{\kappa}$ as follows.

$$\begin{aligned} \mathbb{A}[\boldsymbol{n}] : \operatorname{grad}\boldsymbol{\kappa} &= \mathbb{A}[\boldsymbol{n}][\boldsymbol{n}] \cdot \partial_n\boldsymbol{\kappa} + \mathbb{A}[\boldsymbol{n}] : \operatorname{grad}_{\mathcal{S}}\boldsymbol{\kappa}, \\ &= \mathbb{A}[\boldsymbol{n}][\boldsymbol{n}] \cdot \partial_n\boldsymbol{\kappa} + \operatorname{div}_{\mathcal{S}}((\mathbb{A}[\boldsymbol{n}]\mathbf{P}_{\boldsymbol{n}})^{\top}[\boldsymbol{\kappa}]) \\ &\quad - \boldsymbol{\kappa} \cdot \operatorname{div}_{\mathcal{S}}(\mathbb{A}[\boldsymbol{n}]) - 2K\boldsymbol{\kappa} \cdot \mathbb{A}[\boldsymbol{n}][\boldsymbol{n}], \end{aligned} \tag{126}$$

and

$$
\begin{aligned}
\mathbb{A}[\boldsymbol{n}] \,\vdots\, \operatorname{grad}^2 \boldsymbol{\kappa} = {} & \mathbb{A}[\boldsymbol{n}] \,\vdots\, \operatorname{grad}_{\mathcal{S}}^2 \boldsymbol{\kappa} + \mathbb{A}[\boldsymbol{n}][\boldsymbol{n}]\mathbf{K}^{\top} : \operatorname{grad}_{\mathcal{S}} \boldsymbol{\kappa} \\
& + 2\, \mathbb{A}[\boldsymbol{n}][\boldsymbol{n}] : \operatorname{grad}_{\mathcal{S}}(\partial_n \boldsymbol{\kappa}) - \mathbb{A}[\boldsymbol{n}][\mathbf{K}] \cdot \partial_n \boldsymbol{\kappa} \\
& + \partial_n^2 \boldsymbol{\kappa} \cdot \mathbb{A}[\boldsymbol{n}][\boldsymbol{n}][\boldsymbol{n}], \\
= {} & \operatorname{div}_{\mathcal{S}}((\mathbb{A}[\boldsymbol{n}]\mathbf{P}_n)^{\dashv}[\operatorname{grad}_{\mathcal{S}} \boldsymbol{\kappa}]) \\
& - \operatorname{div}_{\mathcal{S}}(((\operatorname{div}_{\mathcal{S}}(\mathbb{A}[\boldsymbol{n}]) + 2K\, \mathbb{A}[\boldsymbol{n}][\boldsymbol{n}])\mathbf{P}_n)^{\top}[\boldsymbol{\kappa}]) \\
& + \operatorname{div}_{\mathcal{S}}(\operatorname{div}_{\mathcal{S}}(\mathbb{A}[\boldsymbol{n}]) + 2K\, \mathbb{A}[\boldsymbol{n}][\boldsymbol{n}]) \cdot \boldsymbol{\kappa} \\
& + 2K((\operatorname{div}_{\mathcal{S}}(\mathbb{A}[\boldsymbol{n}]))[\boldsymbol{n}] + 2K\, \mathbb{A}[\boldsymbol{n}][\boldsymbol{n}][\boldsymbol{n}]) \cdot \boldsymbol{\kappa} \\
& + \operatorname{div}_{\mathcal{S}}(\mathbb{A}[\boldsymbol{n}][\boldsymbol{n}][\mathbf{K}^{\top}[\boldsymbol{\kappa}]]) - \operatorname{div}_{\mathcal{S}}(\mathbb{A}[\boldsymbol{n}][\boldsymbol{n}]\mathbf{K}^{\top}) \cdot \boldsymbol{\kappa} \\
& + 2 \operatorname{div}_{\mathcal{S}}((\mathbb{A}[\boldsymbol{n}][\boldsymbol{n}]\mathbf{P}_n)^{\top}[\partial_n \boldsymbol{\kappa}]) - 2 \operatorname{div}_{\mathcal{S}}(\mathbb{A}[\boldsymbol{n}][\boldsymbol{n}]) \cdot \partial_n \boldsymbol{\kappa} \\
& - 4K\, \mathbb{A}[\boldsymbol{n}][\boldsymbol{n}][\boldsymbol{n}] \cdot \partial_n \boldsymbol{\kappa} - \mathbb{A}[\boldsymbol{n}][\mathbf{K}] \cdot \partial_n \boldsymbol{\kappa} + \mathbb{A}[\boldsymbol{n}][\boldsymbol{n}][\boldsymbol{n}] \cdot \partial_n^2 \boldsymbol{\kappa}.
\end{aligned}
\tag{127}
$$

Now, replacing identities (122)–(124) into (104) and then applying the divergence theorem, we are again in a position to use identities (126) with $\mathbb{A}[\boldsymbol{n}] := (\mathbb{T} - \operatorname{div} \mathbf{T})[\boldsymbol{n}]$ and (127) with $\mathbb{A}[\boldsymbol{n}] := \mathbf{T}[\boldsymbol{n}]$. Next, using the surface divergence theorem (75) for nonsmooth closed surfaces followed by the gradient theorem $(82)_1$ for nonsmooth closed curves, we arrive at

$$
\begin{aligned}
& \int_{\overline{\mathcal{P}}} \big(\mathbf{T} : \operatorname{grad} \boldsymbol{\chi} + \mathbb{T} \,\vdots\, \operatorname{grad}^2 \boldsymbol{\chi} + \mathbf{T} :: \operatorname{grad}^3 \boldsymbol{\chi}\big)\, \mathrm{d}v \\
& \qquad = \int_{\overline{\mathcal{P}}} \boldsymbol{\chi} \cdot (-\operatorname{div} \mathbf{T} + \operatorname{div}^2 \mathbb{T} - \operatorname{div}^3 \mathbf{T})\, \mathrm{d}v \\
& \qquad\quad + \int_{\partial\overline{\mathcal{P}}} \big(\bar{\boldsymbol{t}}_{\mathcal{S}} \cdot \boldsymbol{\chi} + {}^{2}\bar{\boldsymbol{t}}_{\mathcal{S}} \cdot \partial_n \boldsymbol{\chi} + {}^{3}\bar{\boldsymbol{t}}_{\mathcal{S}} \cdot \partial_n^2 \boldsymbol{\chi}\big)\, \mathrm{d}a \\
& \qquad\quad + \int_{\overline{\mathcal{C}}} \big(\bar{\boldsymbol{t}}_{\mathcal{C}} \cdot \boldsymbol{\chi} + {}^{2}_{n}\bar{\boldsymbol{t}}_{\mathcal{C}} \cdot \partial_n \boldsymbol{\chi} + {}^{2}_{\nu}\bar{\boldsymbol{t}}_{\mathcal{C}} \cdot \partial_\nu \boldsymbol{\chi}\big)\, \mathrm{d}\sigma + \bar{\boldsymbol{t}}_{\mathcal{O}} \cdot \boldsymbol{\chi}|_{\mathcal{O}},
\end{aligned}
\tag{128}
$$

where $\bar{\boldsymbol{t}}_{\mathcal{S}}$, ${}^{2}\bar{\boldsymbol{t}}_{\mathcal{S}}$, ${}^{3}\bar{\boldsymbol{t}}_{\mathcal{S}}$, $\bar{\boldsymbol{t}}_{\mathcal{C}}$, ${}^{2}_{n}\bar{\boldsymbol{t}}_{\mathcal{C}}$, ${}^{2}_{\nu}\bar{\boldsymbol{t}}_{\mathcal{C}}$, and $\bar{\boldsymbol{t}}_{\mathcal{O}}$ are given as follows.

$$
\begin{aligned}
\bar{\boldsymbol{t}}_{\mathcal{S}}(\boldsymbol{x}, t; \boldsymbol{n}, \mathbf{K}, \mathbb{K}) := (\mathbf{T} - \operatorname{div}\mathbb{T} + \operatorname{div}^2\mathbf{T})[\boldsymbol{n}] \\
- \operatorname{div}_{\mathcal{S}}((\mathbb{T} - \operatorname{div}\mathbf{T})[\boldsymbol{n}]) - 2K(\mathbb{T} - \operatorname{div}\mathbf{T})[\boldsymbol{n}][\boldsymbol{n}] \\
+ \operatorname{div}_{\mathcal{S}}(\operatorname{div}_{\mathcal{S}}(\mathbf{T}[\boldsymbol{n}]) + 2K\,\mathbf{T}[\boldsymbol{n}][\boldsymbol{n}]) \\
+ 2K((\operatorname{div}_{\mathcal{S}}(\mathbf{T}[\boldsymbol{n}]))[\boldsymbol{n}] + 2K\,\mathbf{T}[\boldsymbol{n}][\boldsymbol{n}][\boldsymbol{n}]) \\
- \operatorname{div}_{\mathcal{S}}(\mathbf{T}[\boldsymbol{n}][\boldsymbol{n}]\mathbf{K}^{\top}),
\end{aligned}
\tag{129}
$$

$$
\begin{aligned}
{}^{2}\bar{\boldsymbol{t}}_{\mathcal{S}}(\boldsymbol{x}, t; \boldsymbol{n}, \mathbf{K}) := (\mathbb{T} - \operatorname{div}\mathbf{T})[\boldsymbol{n}][\boldsymbol{n}] - 2\operatorname{div}_{\mathcal{S}}(\mathbf{T}[\boldsymbol{n}][\boldsymbol{n}]) \\
- 4K\,\mathbf{T}[\boldsymbol{n}][\boldsymbol{n}][\boldsymbol{n}] - \mathbf{T}[\boldsymbol{n}][\mathbf{K}],
\end{aligned}
\tag{130}
$$

$$
{}^{3}\bar{\boldsymbol{t}}_{\mathcal{S}}(\boldsymbol{x}, t; \boldsymbol{n}) := \mathbf{T}[\boldsymbol{n}][\boldsymbol{n}][\boldsymbol{n}],
\tag{131}
$$

$$
\begin{aligned}
\bar{\boldsymbol{t}}_{\mathcal{C}}(\boldsymbol{x}, t; \boldsymbol{n}^{\pm}, \mathbf{K}^{\pm}) := \{\!\{(\mathbb{T} - \operatorname{div}\mathbf{T})[\boldsymbol{n}][\boldsymbol{\nu}] - \operatorname{div}_{\mathcal{S}}(\mathbf{T}[\boldsymbol{n}])[\boldsymbol{\nu}] \\
- 2K\,\mathbf{T}[\boldsymbol{n}][\boldsymbol{n}][\boldsymbol{\nu}] + \mathbf{T}[\boldsymbol{n}][\boldsymbol{n}][\mathbf{K}^{\top}[\boldsymbol{\nu}]] \\
- \partial_{\sigma}(\mathbf{T}[\boldsymbol{n}][\boldsymbol{\nu}][\boldsymbol{\sigma}])\}\!\},
\end{aligned}
\tag{132}
$$

$$
{}^{2}_{n}\bar{\boldsymbol{t}}_{\mathcal{C}}(\boldsymbol{x}, t; \boldsymbol{n}^{\pm}) := \{\!\{2\,\mathbf{T}[\boldsymbol{n}][\boldsymbol{n}][\boldsymbol{\nu}]\}\!\},
\tag{133}
$$

$$
{}^{2}_{\nu}\bar{\boldsymbol{t}}_{\mathcal{C}}(\boldsymbol{x}, t; \boldsymbol{n}^{\pm}) := \{\!\{\mathbf{T}[\boldsymbol{n}][\boldsymbol{\nu}][\boldsymbol{\nu}]\}\!\},
\tag{134}
$$

and

$$
\bar{\boldsymbol{t}}_{\mathcal{O}}(\boldsymbol{x}, t; {}^{\pm}\boldsymbol{n}^{\pm}) := [\![\{\!\{\mathbf{T}[\boldsymbol{n}][\boldsymbol{\nu}][\boldsymbol{\sigma}]\}\!\}]\!].
\tag{135}
$$

Thus, we have that the principle of virtual powers (101) is satisfied for any arbitrary control volume $\overline{\mathcal{P}} \subset \mathcal{B}$, time t, and any choice of smooth admissible vector $\boldsymbol{\chi}$, if and only if

$$
\begin{aligned}
&\int_{\overline{\mathcal{P}}} \boldsymbol{\chi} \cdot (\boldsymbol{b} + \operatorname{div}\mathbf{T} - \operatorname{div}^2\mathbb{T} + \operatorname{div}^3\mathbf{T})\,\mathrm{d}v \\
&\quad + \int_{\partial\overline{\mathcal{P}}} \big((\boldsymbol{t}_{\mathcal{S}} - \bar{\boldsymbol{t}}_{\mathcal{S}}) \cdot \boldsymbol{\chi} + ({}^{2}\boldsymbol{t}_{\mathcal{S}} - {}^{2}\bar{\boldsymbol{t}}_{\mathcal{S}}) \cdot \partial_n \boldsymbol{\chi} + ({}^{2}\boldsymbol{t}_{\mathcal{S}} - {}^{3}\bar{\boldsymbol{t}}_{\mathcal{S}}) \cdot \partial_n^2 \boldsymbol{\chi}\big)\,\mathrm{d}a \\
&\quad + \int_{\overline{\mathcal{C}}} \big((\boldsymbol{t}_{\mathcal{C}} - \bar{\boldsymbol{t}}_{\mathcal{C}}) \cdot \boldsymbol{\chi} + ({}^{2}_{n}\boldsymbol{t}_{\mathcal{C}} - {}^{2}_{n}\bar{\boldsymbol{t}}_{\mathcal{C}}) \cdot \partial_n \boldsymbol{\chi} + ({}^{2}_{\nu}\boldsymbol{t}_{\mathcal{C}} - {}^{2}_{\nu}\bar{\boldsymbol{t}}_{\mathcal{C}}) \cdot \partial_\nu \boldsymbol{\chi}\big)\,\mathrm{d}\sigma \\
&\qquad\qquad + (\boldsymbol{t}_{\mathcal{O}} - \bar{\boldsymbol{t}}_{\mathcal{O}}) \cdot \boldsymbol{\chi}|_{\mathcal{O}} = \mathbf{0},
\end{aligned}
\tag{136}
$$

which results from combining (128) with (105) through (101). Also, given the arbitrariness and independence among the kinematic processes $\boldsymbol{\chi}$, $\partial_n\boldsymbol{\chi}$, $\partial_n^2\boldsymbol{\chi}$, and $\partial_\nu\boldsymbol{\chi}$, we have that $\boldsymbol{t}_{\mathcal{S}} := \bar{\boldsymbol{t}}_{\mathcal{S}}$, ${}^2\boldsymbol{t}_{\mathcal{S}} := {}^2\bar{\boldsymbol{t}}_{\mathcal{S}}$, ${}^3\boldsymbol{t}_{\mathcal{S}} := {}^3\bar{\boldsymbol{t}}_{\mathcal{S}}$, $\boldsymbol{t}_{\mathcal{C}} := \bar{\boldsymbol{t}}_{\mathcal{C}}$, ${}^2_n\boldsymbol{t}_{\mathcal{C}} := {}^2_n\bar{\boldsymbol{t}}_{\mathcal{C}}$, ${}^2_\nu\boldsymbol{t}_{\mathcal{C}} := {}^2_\nu\bar{\boldsymbol{t}}_{\mathcal{C}}$, and $\boldsymbol{t}_{\mathcal{O}} := \bar{\boldsymbol{t}}_{\mathcal{O}}$, and that the pointwise field equation of forces is given by

$$\operatorname{div}(\mathbf{T} - \operatorname{div}(\mathbb{T} - \operatorname{div}\mathbf{\mathbb{T}})) + \boldsymbol{b} = \mathbf{0}. \tag{137}$$

Reversing arguments, we may also depart from (136) to arrive at (101); then, concluding our proof.

The αth surface microtraction (113), the αth surface hypermicrotraction (114), and the αth junction-edge microtraction (116) may also assume a representation in terms of their tangential components, that is,

$$\begin{aligned}\xi^\alpha_{\mathcal{S}}(\boldsymbol{x}, t; \boldsymbol{n}, \mathbf{K}, \mathbb{K}) := (\boldsymbol{\xi}^\alpha - \operatorname{div}\Xi^\alpha + \operatorname{div}^2\mathbb{X}^\alpha)\cdot\boldsymbol{n} - \operatorname{div}_{\mathcal{S}}(\mathbf{P}_n[(\Xi^\alpha - \operatorname{div}\mathbb{X}^\alpha)[\boldsymbol{n}]])\\ + \operatorname{div}_{\mathcal{S}}(\mathbf{P}_n[\operatorname{div}_{\mathcal{S}}(\mathbb{X}^\alpha[\boldsymbol{n}]\mathbf{P}_n)]) - \operatorname{div}_{\mathcal{S}}(\mathbb{X}^\alpha[\boldsymbol{n}][\boldsymbol{n}]\cdot\mathbf{K}^\top),\end{aligned} \tag{138}$$

$${}^2\xi^\alpha_{\mathcal{S}}(\boldsymbol{x}, t; \boldsymbol{n}, \mathbf{K}) := \boldsymbol{n}\cdot(\Xi^\alpha - \operatorname{div}\mathbb{X}^\alpha)[\boldsymbol{n}] - 2\operatorname{div}_{\mathcal{S}}(\mathbf{P}_n[\mathbb{X}^\alpha[\boldsymbol{n}][\boldsymbol{n}]]) - \mathbb{X}^\alpha[\boldsymbol{n}] : \mathbf{K}, \tag{139}$$

and

$$\begin{aligned}\xi^\alpha_{\mathcal{C}}(\boldsymbol{x}, t; \boldsymbol{n}^\pm, \mathbf{K}^\pm) := \{\!\!\{\boldsymbol{\nu}\cdot((\Xi^\alpha - \operatorname{div}\mathbb{X}^\alpha)[\boldsymbol{n}] - \operatorname{div}_{\mathcal{S}}(\mathbb{X}^\alpha[\boldsymbol{n}]\mathbf{P}_n)\\ + \mathbb{X}^\alpha[\boldsymbol{n}][\boldsymbol{n}]\cdot\mathbf{K}^\top) - \partial_\sigma(\boldsymbol{\sigma}\cdot\mathbb{X}^\alpha[\boldsymbol{n}][\boldsymbol{\nu}])\}\!\!\}.\end{aligned} \tag{140}$$

Analogously, the surface traction (129), the surface hypertraction (130), and the junction-edge traction (132) may also assume a representation in terms of their tangential components, that is,

$$\begin{aligned}\boldsymbol{t}_{\mathcal{S}}(\boldsymbol{x}, t; \boldsymbol{n}, \mathbf{K}, \mathbb{K}) := (\mathbf{T} - \operatorname{div}\mathbb{T} + \operatorname{div}^2\mathbf{\mathbb{T}})[\boldsymbol{n}] - \operatorname{div}_{\mathcal{S}}((\mathbb{T} - \operatorname{div}\mathbf{\mathbb{T}})[\boldsymbol{n}]\mathbf{P}_n)\\ + \operatorname{div}_{\mathcal{S}}(\operatorname{div}_{\mathcal{S}}(\mathbf{\mathbb{T}}[\boldsymbol{n}]\mathbf{P}_n)\mathbf{P}_n) - \operatorname{div}_{\mathcal{S}}(\mathbf{\mathbb{T}}[\boldsymbol{n}][\boldsymbol{n}]\mathbf{K}^\top),\end{aligned} \tag{141}$$

$${}^2\boldsymbol{t}_{\mathcal{S}}(\boldsymbol{x}, t; \boldsymbol{n}, \mathbf{K}) := (\mathbb{T} - \operatorname{div}\mathbf{\mathbb{T}})[\boldsymbol{n}][\boldsymbol{n}] - 2\operatorname{div}_{\mathcal{S}}(\mathbf{\mathbb{T}}[\boldsymbol{n}][\boldsymbol{n}]\mathbf{P}_n) - \mathbf{\mathbb{T}}[\boldsymbol{n}][\mathbf{K}]. \tag{142}$$

and

$$\begin{aligned}\boldsymbol{t}_{\mathcal{C}}(\boldsymbol{x}, t; \boldsymbol{n}^\pm, \mathbf{K}^\pm) := \{\!\!\{(\mathbb{T} - \operatorname{div}\mathbf{\mathbb{T}})[\boldsymbol{n}][\boldsymbol{\nu}] - \operatorname{div}_{\mathcal{S}}(\mathbf{\mathbb{T}}[\boldsymbol{n}]\mathbf{P}_n)[\boldsymbol{\nu}]\\ + \mathbf{\mathbb{T}}[\boldsymbol{n}][\boldsymbol{n}][\mathbf{K}^\top[\boldsymbol{\nu}]] - \partial_\sigma(\mathbf{\mathbb{T}}[\boldsymbol{n}][\boldsymbol{\nu}][\boldsymbol{\sigma}])\}\!\!\}.\end{aligned} \tag{143}$$

Letting the phase fields be denoted by $\{\varphi^\alpha\}_{\alpha=1}^n$ and the velocity by $\boldsymbol{v}$, the actual external power expenditures are given by

$$\mathcal{W}^{\flat}_{\text{ext}}(\overline{\mathcal{P}}) := \mathcal{V}^{\flat}_{\text{ext}}(\overline{\mathcal{P}}; \{\dot{\varphi}^{\alpha}\}^{n}_{\alpha=1}) \quad \text{and} \quad \mathcal{W}^{\sharp}_{\text{ext}}(\overline{\mathcal{P}}) := \mathcal{V}^{\sharp}_{\text{ext}}(\overline{\mathcal{P}}; \boldsymbol{v}), \tag{144}$$

while the internal power expenditures are

$$\mathcal{W}^{\flat}_{\text{int}}(\overline{\mathcal{P}}) := \mathcal{V}^{\flat}_{\text{int}}(\overline{\mathcal{P}}; \{\dot{\varphi}^{\alpha}\}^{n}_{\alpha=1}) \quad \text{and} \quad \mathcal{W}^{\sharp}_{\text{int}}(\overline{\mathcal{P}}) := \mathcal{V}^{\sharp}_{\text{int}}(\overline{\mathcal{P}}; \boldsymbol{v}). \tag{145}$$

Thus, the total power expenditures are

$$\mathcal{W}_{\text{int}}(\overline{\mathcal{P}}) := \mathcal{W}^{\flat}_{\text{int}}(\overline{\mathcal{P}}) + \mathcal{W}^{\sharp}_{\text{int}}(\overline{\mathcal{P}}) \quad \text{and} \quad \mathcal{W}_{\text{ext}}(\overline{\mathcal{P}}) := \mathcal{W}^{\flat}_{\text{ext}}(\overline{\mathcal{P}}) + \mathcal{W}^{\sharp}_{\text{ext}}(\overline{\mathcal{P}}). \tag{146}$$

Lastly, the actual power balance is

$$\mathcal{W}_{\text{ext}}(\overline{\mathcal{P}}) = \mathcal{W}_{\text{int}}(\overline{\mathcal{P}}). \tag{147}$$

As we mentioned in the introductory part of this study, the Euler–Cauchy hypothesis, which establishes that the surface microtraction and the surface traction depend upon the unit normal field of the Euler–Cauchy cut, is abandoned. In the following remark, we motivate the dependency upon the geometrical descriptors $\{\boldsymbol{n}, \mathbf{K}, \mathbb{K}\}$.

Remark 1 (The αth surface microtraction $\xi^{\alpha}_{\mathcal{S}}$ and the surface traction $\boldsymbol{t}_{\mathcal{S}}$ as function of the normal $\boldsymbol{n}$, the curvature tensor $\mathbf{K}$, and the second curvature gradient $\mathbb{K}$) For the point we wish to make, it suffices to study one term. In the αth surface microtraction $\xi^{\alpha}_{\mathcal{S}}$ in expression (113) and the surface traction $\boldsymbol{t}_{\mathcal{S}}$ (129), we have, respectively, the following terms

$$\begin{aligned}
\operatorname{div}_{\mathcal{S}}(\operatorname{div}_{\mathcal{S}}(\mathbb{X}^{\alpha}[\boldsymbol{n}])) &= \operatorname{div}_{\mathcal{S}}((\operatorname{div}_{\mathcal{S}}(\mathbb{X}^{\alpha})^{\top})[\boldsymbol{n}] - \mathbb{X}^{\alpha}[\mathbf{K}]), \\
&= (\operatorname{div}^2_{\mathcal{S}}(\mathbb{X}^{\alpha})^{\dashv})[\boldsymbol{n}] - (\operatorname{div}_{\mathcal{S}}(\mathbb{X}^{\alpha})^{\top}) : \mathbf{K} \\
&\quad - \operatorname{div}_{\mathcal{S}}(\mathbb{X}^{\alpha})^{\vdash} : \mathbf{K} - \mathbb{X}^{\alpha} : \mathbb{K}, \\
&= (\operatorname{div}^2_{\mathcal{S}}\mathbb{X}^{\alpha})[\boldsymbol{n}] - 2\,(\operatorname{div}_{\mathcal{S}}\mathbb{X}^{\alpha}) : \mathbf{K} - \mathbb{X}^{\alpha} : \mathbb{K},
\end{aligned} \tag{148}$$

and

$$\begin{aligned}
\operatorname{div}_{\mathcal{S}}(\operatorname{div}_{\mathcal{S}}(\mathbb{T}[\boldsymbol{n}])) &= \operatorname{div}_{\mathcal{S}}((\operatorname{div}_{\mathcal{S}}\mathbb{T}^{\top})[\boldsymbol{n}] - \mathbb{T}[\mathbf{K}]), \\
&= (\operatorname{div}^2_{\mathcal{S}}\mathbb{T}^{\dashv})^{\top}[\boldsymbol{n}] - (\operatorname{div}_{\mathcal{S}}\mathbb{T}^{\top})[\mathbf{K}] \\
&\quad - (\operatorname{div}_{\mathcal{S}}(\mathbb{T}^{\dashv})^{\dashv})^{\dashv}[\mathbf{K}] - \mathbb{T}[\mathbb{K}], \\
&= (\operatorname{div}^2_{\mathcal{S}}\mathbb{T})[\boldsymbol{n}] - 2\,(\operatorname{div}_{\mathcal{S}}\mathbb{T})[\mathbf{K}] - \mathbb{T}[\mathbb{K}].
\end{aligned} \tag{149}$$

Each term represents the dependency on tuple $\{\boldsymbol{n}, \mathbf{K}, \mathbb{K}\}$.

Although we have described the junction-edge microtraction and junction-edge traction depending upon the limiting unit normal field $\boldsymbol{n}^{\pm}$ and the limiting curvature tensor field $\mathbf{K}^{\pm}$, there is one term that may be expressed in terms of the normal curvature, the geodesic curvature, and the geodesic torsion of the junction-edge where these microtractions and tractions are developed. Moreover, this term connects the junction-edge microtraction with the junction-point microforce, and analogously, the junction-edge traction with the junction-point force, through the gradient theorem on nonsmooth curves $(82)_1$. We make explicit this dependency in the following remark.

Remark 2 ($\{\xi_{\mathcal{C}}^{\alpha}, \boldsymbol{t}_{\mathcal{C}}\}$ and their dependency on the normal curvature, the geodesic curvature, and the geodesic torsion of $\mathcal{C}$) Recalling the geodesic curvatures and the geodesic torsion in (66), we have that the last terms in (116) and (132) may assume the following form

$$
\begin{aligned}
\partial_\sigma(\boldsymbol{\sigma} \cdot \mathbb{X}^\alpha[\boldsymbol{n}][\boldsymbol{\nu}]) &= \boldsymbol{\sigma} \cdot (\partial_\sigma \mathbb{X}^\alpha)[\boldsymbol{n}][\boldsymbol{\nu}] \\
&\quad - k_n \boldsymbol{\sigma} \cdot \mathbb{X}^\alpha[\boldsymbol{\sigma}][\boldsymbol{\nu}] - t_g \boldsymbol{\sigma} \cdot \mathbb{X}^\alpha[\boldsymbol{\nu}][\boldsymbol{\nu}] \\
&\quad + k_g \boldsymbol{\sigma} \cdot \mathbb{X}^\alpha[\boldsymbol{n}][\boldsymbol{\sigma}] + t_g \boldsymbol{\sigma} \cdot \mathbb{X}^\alpha[\boldsymbol{n}][\boldsymbol{n}] \\
&\quad + k_n \boldsymbol{n} \cdot \mathbb{X}^\alpha[\boldsymbol{n}][\boldsymbol{\nu}] - k_g \boldsymbol{\nu} \cdot \mathbb{X}^\alpha[\boldsymbol{n}][\boldsymbol{\nu}], \\
&= \boldsymbol{\sigma} \cdot (\partial_\sigma \mathbb{X}^\alpha)[\boldsymbol{n}][\boldsymbol{\nu}] \\
&\quad + k_n \mathbb{X}^\alpha[\boldsymbol{\nu}] : (\boldsymbol{n} \otimes \boldsymbol{n} - \boldsymbol{\sigma} \otimes \boldsymbol{\sigma}) \\
&\quad + k_g \mathbb{X}^\alpha[\boldsymbol{n}] : (\boldsymbol{\sigma} \otimes \boldsymbol{\sigma} - \boldsymbol{\nu} \otimes \boldsymbol{\nu}) \\
&\quad + t_g \mathbb{X}^\alpha[\boldsymbol{\sigma}] : (\boldsymbol{n} \otimes \boldsymbol{n} - \boldsymbol{\nu} \otimes \boldsymbol{\nu}),
\end{aligned} \tag{150}
$$

and

$$
\begin{aligned}
\partial_\sigma(\mathbb{T}[\boldsymbol{n}][\boldsymbol{\nu}][\boldsymbol{\sigma}]) &= (\partial_\sigma \mathbb{T})[\boldsymbol{n}][\boldsymbol{\nu}][\boldsymbol{\sigma}] \\
&\quad - k_n \mathbb{T}[\boldsymbol{\sigma}][\boldsymbol{\nu}][\boldsymbol{\sigma}] - t_g \mathbb{T}[\boldsymbol{\nu}][\boldsymbol{\nu}][\boldsymbol{\sigma}] \\
&\quad + k_g \mathbb{T}[\boldsymbol{n}][\boldsymbol{\sigma}][\boldsymbol{\sigma}] + t_g \mathbb{T}[\boldsymbol{n}][\boldsymbol{n}][\boldsymbol{\sigma}] \\
&\quad + k_n \mathbb{T}[\boldsymbol{n}][\boldsymbol{\nu}][\boldsymbol{n}] - k_g \mathbb{T}[\boldsymbol{n}][\boldsymbol{\nu}][\boldsymbol{\nu}], \\
&= (\partial_\sigma \mathbb{T})[\boldsymbol{n}][\boldsymbol{\nu}][\boldsymbol{\sigma}] \\
&\quad + k_n \mathbb{T}[\boldsymbol{\nu}][\boldsymbol{n} \otimes \boldsymbol{n} - \boldsymbol{\sigma} \otimes \boldsymbol{\sigma}] \\
&\quad + k_g \mathbb{T}[\boldsymbol{n}][\boldsymbol{\sigma} \otimes \boldsymbol{\sigma} - \boldsymbol{\nu} \otimes \boldsymbol{\nu}] \\
&\quad + t_g \mathbb{T}[\boldsymbol{\sigma}][\boldsymbol{n} \otimes \boldsymbol{n} - \boldsymbol{\nu} \otimes \boldsymbol{\nu}].
\end{aligned} \tag{151}
$$

Note that we have not used the *surplus* notation, $\{\!\!\{\cdot\}\!\!\}$ for the sake of brevity. In fact, the normal curvature, the geodesic curvature, and the geodesic torsion appearing in these terms are the limiting ones, that is, $k_n^{\pm}$, $k_g^{\pm}$, and $t_g^{\pm}$.

2 Inertia Components

For convenience, we may rewrite the external actual power (105) for a spatial part $\mathcal{P}_\tau$ advecting with the body $\mathcal{B}_\tau$ with $\boldsymbol{\chi} := \boldsymbol{v}$ and using the initial and noninertial contributions of $\boldsymbol{b} = \boldsymbol{b}^{\mathrm{ni}} - \varrho\dot{\boldsymbol{v}}$, that is,

$$\mathcal{W}^{\sharp}_{\mathrm{ext}}(\mathcal{P}_\tau) = -\tfrac{1}{2}\overline{\int_{\mathcal{P}_\tau} \varrho|\boldsymbol{v}|^2 \,\mathrm{d}v_\tau}^{\,\boldsymbol{\cdot}} + \int_{\mathcal{P}_\tau} \boldsymbol{b}^{\mathrm{ni}} \cdot \boldsymbol{v} \,\mathrm{d}v_\tau$$
$$+ \int_{\partial\mathcal{P}_\tau} \big(\boldsymbol{t}_{\mathcal{S}} \cdot \boldsymbol{v} + {}^{2}\boldsymbol{t}_{\mathcal{S}} \cdot \partial_n \boldsymbol{v} + {}^{3}\boldsymbol{t}_{\mathcal{S}} \cdot \partial_n^2 \boldsymbol{v}\big)\,\mathrm{d}a_\tau$$
$$+ \int_{\mathcal{C}_\tau} \big(\boldsymbol{t}_{\mathcal{C}} \cdot \boldsymbol{v} + {}^{2}_{n}\boldsymbol{t}_{\mathcal{C}} \cdot \partial_n \boldsymbol{v} + {}^{2}_{\nu}\boldsymbol{t}_{\mathcal{C}} \cdot \partial_\nu \boldsymbol{v}\big)\,\mathrm{d}\sigma_\tau + \boldsymbol{t}_{\mathcal{O}} \cdot \boldsymbol{v}|_{\mathcal{O}}. \quad (152)$$

Synopsis, chapter 3

Microkinetics

(i) Partwise balance of virtual powers on $\overline{\mathcal{P}}$ for scalar virtual fields:

$$\underbrace{\sum_{\alpha=1}^{n}\int_{\overline{\mathcal{P}}}\big(-\pi^\alpha\chi^\alpha+\boldsymbol{\xi}^\alpha\cdot\operatorname{grad}\chi^\alpha+\boldsymbol{\Xi}^\alpha:\operatorname{grad}^2\chi^\alpha+\mathbb{X}^\alpha:\operatorname{grad}^3\chi^\alpha\big)\,\mathrm{d}v}_{\mathcal{V}^\flat_{\mathrm{int}}(\overline{\mathcal{P}};\{\chi^\alpha\}_{\alpha=1}^n)}=$$

$$\underbrace{\sum_{\alpha=1}^{n}\Big\{\int_{\overline{\mathcal{P}}}\gamma^\alpha\chi^\alpha\,\mathrm{d}v+\int_{\partial\overline{\mathcal{P}}}\big(\xi^\alpha_{\mathcal{S}}\chi^\alpha+\xi^\alpha_{\mathcal{S}}\partial_n\chi^\alpha+{}^3\xi^\alpha_{\mathcal{S}}\partial^2_n\chi^\alpha\big)\,\mathrm{d}a+\int_{\overline{\mathcal{C}}}\big(\xi^\alpha_{\mathcal{C}}\chi^\alpha+{}^2_n\xi^\alpha_{\mathcal{C}}\partial_n\chi^\alpha+{}^2_\nu\xi^\alpha_{\mathcal{C}}\partial_\nu\chi^\alpha\big)\,\mathrm{d}\sigma+\xi^\alpha_{\mathcal{O}}\chi^\alpha|_{\mathcal{O}}\Big\}}_{\mathcal{V}^\flat_{\mathrm{ext}}(\overline{\mathcal{P}};\{\chi^\alpha\}_{\alpha=1}^n)}$$

(ii) Pointwise field equation of microforces:

$$\operatorname{div}(\boldsymbol{\xi}^\alpha-\operatorname{div}(\boldsymbol{\Xi}^\alpha-\operatorname{div}\mathbb{X}^\alpha))+\pi^\alpha+\gamma^\alpha=0\qquad\forall\,1\le\alpha\le n$$

(iii) Surface microtraction on $\partial\overline{\mathcal{P}}$:

$$\begin{aligned}
{}^2\xi^\alpha_{\mathcal{S}}(\boldsymbol{x},t;\boldsymbol{n},\mathbf{K},\mathbb{K}) := {} & (\boldsymbol{\xi}^\alpha-\boldsymbol{\Xi}^\alpha+\operatorname{div}^2\mathbb{X}^\alpha)\cdot\boldsymbol{n}\\
& -\operatorname{div}_{\mathcal{S}}((\boldsymbol{\Xi}^\alpha-\operatorname{div}\mathbb{X}^\alpha)[\boldsymbol{n}])-2K\,\boldsymbol{n}\cdot(\boldsymbol{\Xi}^\alpha-\operatorname{div}\mathbb{X}^\alpha)[\boldsymbol{n}]\\
& +\operatorname{div}_{\mathcal{S}}(\operatorname{div}_{\mathcal{S}}(\mathbb{X}^\alpha[\boldsymbol{n}])+2K\,\mathbb{X}^\alpha[\boldsymbol{n}][\boldsymbol{n}])\\
& +2K\,\boldsymbol{n}\cdot(\operatorname{div}_{\mathcal{S}}(\mathbb{X}^\alpha[\boldsymbol{n}])+2K\,\mathbb{X}^\alpha[\boldsymbol{n}][\boldsymbol{n}])\\
& -\operatorname{div}_{\mathcal{S}}(\mathbb{X}^\alpha[\boldsymbol{n}][\boldsymbol{n}]\cdot\mathbf{K}^{\top})\\
= {} & (\boldsymbol{\xi}^\alpha-\operatorname{div}\boldsymbol{\Xi}^\alpha+\operatorname{div}^2\mathbb{X}^\alpha)\cdot\boldsymbol{n}\\
& -\operatorname{div}_{\mathcal{S}}(\mathbf{P}_n[(\boldsymbol{\Xi}^\alpha-\operatorname{div}\mathbb{X}^\alpha)[\boldsymbol{n}]])\\
& +\operatorname{div}_{\mathcal{S}}(\mathbf{P}_n[\operatorname{div}_{\mathcal{S}}(\mathbb{X}^\alpha[\boldsymbol{n}]\mathbf{P}_n)])\\
& -\operatorname{div}_{\mathcal{S}}(\mathbb{X}^\alpha[\boldsymbol{n}][\boldsymbol{n}]\cdot\mathbf{K}^{\top})
\end{aligned}$$

(iv) *Surface hypermicrotraction on $\overline{\partial\mathcal{P}}$:*

$$
\begin{aligned}
{}^{2}\xi^{\alpha}_{\mathcal{S}}(\boldsymbol{x},t;\boldsymbol{n},\mathbf{K}) &:= \boldsymbol{n}\cdot(\boldsymbol{\Xi}^{\alpha}-\operatorname{div}\mathbb{X}^{\alpha})[\boldsymbol{n}] - 2\operatorname{div}_{\mathcal{S}}(\mathbb{X}^{\alpha}[\boldsymbol{n}][\boldsymbol{n}]) \\
&\quad - 4K\,\boldsymbol{n}\cdot\mathbb{X}^{\alpha}[\boldsymbol{n}][\boldsymbol{n}] - \mathbb{X}^{\alpha}[\boldsymbol{n}]:\mathbf{K} \\
&= \boldsymbol{n}\cdot(\boldsymbol{\Xi}^{\alpha}-\operatorname{div}\mathbb{X}^{\alpha})[\boldsymbol{n}] - 2\operatorname{div}_{\mathcal{S}}(\mathbf{P}_{\!n}[\mathbb{X}^{\alpha}[\boldsymbol{n}][\boldsymbol{n}]]) - \mathbb{X}^{\alpha}[\boldsymbol{n}]:\mathbf{K}
\end{aligned}
$$

(v) *Surface supra-hypermicrotraction on $\overline{\partial\mathcal{P}}$:*

$$
{}^{3}\xi^{\alpha}_{\mathcal{S}}(\boldsymbol{x},t;\boldsymbol{n}) := \boldsymbol{n}\cdot\mathbb{X}^{\alpha}[\boldsymbol{n}][\boldsymbol{n}]
$$

(vi) *Junction-edge microtraction on $\overline{\mathcal{C}}$:*

$$
\begin{aligned}
\xi^{\alpha}_{\mathcal{C}}(\boldsymbol{x},t;\boldsymbol{n}^{\pm},\mathbf{K}^{\pm}) &:= \{\!\{\boldsymbol{\nu}\cdot((\boldsymbol{\Xi}^{\alpha}-\operatorname{div}\mathbb{X}^{\alpha})[\boldsymbol{n}] - \operatorname{div}_{\mathcal{S}}(\mathbb{X}^{\alpha}[\boldsymbol{n}]) \\
&\quad - 2K\,\mathbb{X}^{\alpha}[\boldsymbol{n}][\boldsymbol{n}] + \mathbb{X}^{\alpha}[\boldsymbol{n}][\boldsymbol{n}]\cdot\mathbf{K}^{\top}) \\
&\quad - \partial_{\sigma}(\boldsymbol{\sigma}\cdot\mathbb{X}^{\alpha}[\boldsymbol{n}][\boldsymbol{\nu}])\}\!\} \\
&= \{\!\{\boldsymbol{\nu}\cdot((\boldsymbol{\Xi}^{\alpha}-\operatorname{div}\mathbb{X}^{\alpha})[\boldsymbol{n}] - \operatorname{div}_{\mathcal{S}}(\mathbb{X}^{\alpha}[\boldsymbol{n}]\mathbf{P}_{\!n}) \\
&\quad + \mathbb{X}^{\alpha}[\boldsymbol{n}][\boldsymbol{n}]\cdot\mathbf{K}^{\top}) - \partial_{\sigma}(\boldsymbol{\sigma}\cdot\mathbb{X}^{\alpha}[\boldsymbol{n}][\boldsymbol{\nu}])\}\!\}
\end{aligned}
$$

(vii) *Junction-edge hypermicrotraction on $\overline{\mathcal{C}}$:*

$$
{}^{2}_{n}\xi^{\alpha}_{\mathcal{C}}(\boldsymbol{x},t;\boldsymbol{n}^{\pm}) := \{\!\{2\,\boldsymbol{\nu}\cdot\mathbb{X}^{\alpha}[\boldsymbol{n}][\boldsymbol{n}]\}\!\}
$$

$$
{}^{2}_{\nu}\xi^{\alpha}_{\mathcal{C}}(\boldsymbol{x},t;\boldsymbol{n}^{\pm}) := \{\!\{\boldsymbol{\nu}\cdot\mathbb{X}^{\alpha}[\boldsymbol{n}][\boldsymbol{\nu}]\}\!\}
$$

(viii) *Junction-point microforce on $\overline{\mathcal{O}}$:*

$$
\xi^{\alpha}_{\mathcal{O}}(\boldsymbol{x},t;{}^{\pm}\boldsymbol{n}^{\pm}) := [\![\{\!\{\boldsymbol{\sigma}\cdot\mathbb{X}^{\alpha}[\boldsymbol{n}][\boldsymbol{\nu}]\}\!\}]\!]
$$

Kinetics

(i) *Partwise balance of virtual powers on $\overline{\mathcal{P}}$ for a vector virtual field:*

$$\underbrace{\int_{\overline{\mathcal{P}}} \left(\mathbf{T} : \operatorname{grad} \chi + \mathbb{T} \vdots \operatorname{grad}^2 \chi + \mathbf{T} :: \operatorname{grad}^3 \chi\right) \mathrm{d}v}_{\mathcal{V}^\sharp_{\mathrm{int}}(\overline{\mathcal{P}};\chi)} =$$

$$\underbrace{\begin{aligned} &\int_{\overline{\mathcal{P}}} \boldsymbol{b} \cdot \chi \, \mathrm{d}v + \int_{\partial\overline{\mathcal{P}}} \left(\boldsymbol{t}_{\mathcal{S}} \cdot \chi + {}^2\boldsymbol{t}_{\mathcal{S}} \cdot \partial_n \chi + {}^3\boldsymbol{t}_{\mathcal{S}} \cdot \partial_n^2 \chi\right) \mathrm{d}a \\ &+ \int_{\overline{\mathcal{C}}} \left(\boldsymbol{t}_{\mathcal{C}} \cdot \chi + {}^2_n\boldsymbol{t}_{\mathcal{C}} \cdot \partial_n \chi + {}^2_\nu\boldsymbol{t}_{\mathcal{C}} \cdot \partial_\nu \chi\right) \mathrm{d}\sigma + \boldsymbol{t}_{\mathcal{O}} \cdot \chi|_{\mathcal{O}} \end{aligned}}_{\mathcal{V}^\sharp_{\mathrm{ext}}(\overline{\mathcal{P}};\chi)}$$

(ii) Pointwise field equation of forces:

$$\operatorname{div}(\mathbf{T} - \operatorname{div}(\mathbb{T} - \operatorname{div}\mathbf{T})) + \boldsymbol{b} = \mathbf{0}$$

(iii) Surface traction on $\overline{\partial\mathcal{P}}$:

$$\begin{aligned} \boldsymbol{t}_{\mathcal{S}}(\boldsymbol{x}, t; \boldsymbol{n}, \mathbf{K}, \mathbb{K}) := &(\mathbf{T} - \operatorname{div}\mathbb{T} + \operatorname{div}^2\mathbf{T})[\boldsymbol{n}] \\ &- \operatorname{div}_{\mathcal{S}}((\mathbb{T} - \operatorname{div}\mathbf{T})[\boldsymbol{n}]) - 2K(\mathbb{T} - \operatorname{div}\mathbf{T})[\boldsymbol{n}][\boldsymbol{n}] \\ &+ \operatorname{div}_{\mathcal{S}}(\operatorname{div}_{\mathcal{S}}(\mathbf{T}[\boldsymbol{n}]) + 2K\, \mathbf{T}[\boldsymbol{n}][\boldsymbol{n}]) \\ &+ 2K((\operatorname{div}_{\mathcal{S}}(\mathbf{T}[\boldsymbol{n}]))[\boldsymbol{n}] + 2K\, \mathbf{T}[\boldsymbol{n}][\boldsymbol{n}][\boldsymbol{n}]) \\ &- \operatorname{div}_{\mathcal{S}}(\mathbf{T}[\boldsymbol{n}][\boldsymbol{n}]\mathbf{K}^\top) \\ = &(\mathbf{T} - \operatorname{div}\mathbb{T} + \operatorname{div}^2\mathbf{T})[\boldsymbol{n}] \\ &- \operatorname{div}_{\mathcal{S}}((\mathbb{T} - \operatorname{div}\mathbf{T})[\boldsymbol{n}]\mathbf{P}_n) \\ &+ \operatorname{div}_{\mathcal{S}}(\operatorname{div}_{\mathcal{S}}(\mathbf{T}[\boldsymbol{n}]\mathbf{P}_n)\mathbf{P}_n) \\ &- \operatorname{div}_{\mathcal{S}}(\mathbf{T}[\boldsymbol{n}][\boldsymbol{n}]\mathbf{K}^\top) \end{aligned}$$

(iv) Surface hypertraction on $\overline{\partial\mathcal{P}}$:

$$\begin{aligned} {}^2\boldsymbol{t}_{\mathcal{S}}(\boldsymbol{x}, t; \boldsymbol{n}, \mathbf{K}) := &(\mathbb{T} - \operatorname{div}\mathbf{T})[\boldsymbol{n}][\boldsymbol{n}] - 2\operatorname{div}_{\mathcal{S}}(\mathbf{T}[\boldsymbol{n}][\boldsymbol{n}]) \\ &- 4K\, \mathbf{T}[\boldsymbol{n}][\boldsymbol{n}][\boldsymbol{n}] - \mathbf{T}[\boldsymbol{n}][\mathbf{K}] \\ = &(\mathbb{T} - \operatorname{div}\mathbf{T})[\boldsymbol{n}][\boldsymbol{n}] - 2\operatorname{div}_{\mathcal{S}}(\mathbf{T}[\boldsymbol{n}][\boldsymbol{n}]\mathbf{P}_n) - \mathbf{T}[\boldsymbol{n}][\mathbf{K}] \end{aligned}$$

(v) Surface supra-hypertraction on $\overline{\partial\mathcal{P}}$:

$${}^3\boldsymbol{t}_{\mathcal{S}}(\boldsymbol{x}, t; \boldsymbol{n}) := \mathbf{T}[\boldsymbol{n}][\boldsymbol{n}][\boldsymbol{n}]$$

(vi) Junction-edge traction on $\overline{\mathcal{C}}$:

$$\begin{aligned}\boldsymbol{t}_{\mathcal{C}}(\boldsymbol{x}, t; \boldsymbol{n}^{\pm}, \mathbf{K}^{\pm}) := \{\!\!\{ & (\mathbb{T} - \operatorname{div} \mathbf{T})[\boldsymbol{n}][\boldsymbol{v}] - \operatorname{div}_{\mathcal{S}}(\mathbf{T}[\boldsymbol{n}])[\boldsymbol{v}] \\ & - 2K\, \mathbf{T}[\boldsymbol{n}][\boldsymbol{n}][\boldsymbol{v}] + \mathbf{T}[\boldsymbol{n}][\boldsymbol{n}][\mathbf{K}^{\top}[\boldsymbol{v}]] \\ & - \partial_{\sigma}(\mathbf{T}[\boldsymbol{n}][\boldsymbol{v}][\boldsymbol{\sigma}])\}\!\!\} \\ = \{\!\!\{ & (\mathbb{T} - \operatorname{div} \mathbf{T})[\boldsymbol{n}][\boldsymbol{v}] - \operatorname{div}_{\mathcal{S}}(\mathbf{T}[\boldsymbol{n}]\mathbf{P}_{\!n})[\boldsymbol{v}] \\ & + \mathbf{T}[\boldsymbol{n}][\boldsymbol{n}][\mathbf{K}^{\top}[\boldsymbol{v}]] - \partial_{\sigma}(\mathbf{T}[\boldsymbol{n}][\boldsymbol{v}][\boldsymbol{\sigma}])\}\!\!\}\end{aligned}$$

(vii) Junction-edge hypertraction on $\overline{\mathcal{C}}$:

$$\prescript{2}{n}{\boldsymbol{t}}_{\mathcal{C}}(\boldsymbol{x}, t; \boldsymbol{n}^{\pm}) := \{\!\!\{2\,\mathbf{T}[\boldsymbol{n}][\boldsymbol{n}][\boldsymbol{v}]\}\!\!\}$$

$$\prescript{2}{v}{\boldsymbol{t}}_{\mathcal{C}}(\boldsymbol{x}, t; \boldsymbol{n}^{\pm}) := \{\!\!\{\mathbf{T}[\boldsymbol{n}][\boldsymbol{v}][\boldsymbol{v}]\}\!\!\}$$

(viii) Junction-point force on $\overline{\mathcal{O}}$:

$$\boldsymbol{t}_{\mathcal{O}}(\boldsymbol{x}, t; {}^{\pm}\boldsymbol{n}^{\pm}) := [\![\{\!\!\{\mathbf{T}[\boldsymbol{n}][\boldsymbol{v}][\boldsymbol{\sigma}]\}\!\!\}]\!]$$

Chapter 4
Complementary Balances, Jump Conditions, and Couple-Fields

In this chapter, we postulate and study the consequence of frame indifference. Additionally, we derive the partwise balances of microforces, forces, microtorques, and torques from the field equations (121) and (137). Next, we describe the action-reaction principle. Then, from surface balance postulates, we determine jump conditions across surfaces. Lastly, we conclude the chapter with the principle of virtual powers for arbitrary surfaces. Important to what follows is the vector $\boldsymbol{r} := \boldsymbol{x} - \boldsymbol{o}$, where $\boldsymbol{o}$ is arbitrary and fixed while $\boldsymbol{r}_{\circ} := \boldsymbol{x}_{\circ} - \boldsymbol{o}$.

1 Frame Indifference

Postulate 2 (Frame indifference) *The internal virtual power $\mathcal{V}^{\sharp}_{\text{int}}(\overline{\mathcal{P}}; \chi)$, given in (104), is indifferent under changes in frame. Moreover, changes in the frame may be stated by the following mapping*

$$\chi(\boldsymbol{x}, t) \mapsto \chi(\boldsymbol{x}, t) + \boldsymbol{\beta}(t) + \boldsymbol{\Omega}(t)[\boldsymbol{x}], \tag{153}$$

where $\boldsymbol{\beta}(t)$ is an arbitrary vector and $\boldsymbol{\Omega}(t)$ is a second-order skew-symmetric tensor that can vary in time. Thus,

$$\mathcal{V}^{\sharp}_{\text{int}}(\overline{\mathcal{P}}; \chi) = \mathcal{V}^{\sharp}_{\text{int}}(\overline{\mathcal{P}}; \chi(\boldsymbol{x}, t) + \boldsymbol{\beta}(t) + \boldsymbol{\Omega}(t)[\boldsymbol{x}]). \tag{154}$$

From (153), we also have that

L. Espath, *Mechanics and Geometry of Enriched Continua*,
https://doi.org/10.1007/978-3-031-28934-7_4

$$\begin{cases} \operatorname{grad}\chi(\boldsymbol{x},t) \mapsto \operatorname{grad}\chi(\boldsymbol{x},t) + \boldsymbol{\Omega}(t), \\ \operatorname{grad}^2\chi(\boldsymbol{x},t) \mapsto \operatorname{grad}^2\chi(\boldsymbol{x},t), \\ \operatorname{grad}^3\chi(\boldsymbol{x},t) \mapsto \operatorname{grad}^3\chi(\boldsymbol{x},t). \end{cases} \tag{155}$$

Thus, the requirement arising with (154) is that[1]

$$\int_{\mathcal{P}} \mathbf{T} : \boldsymbol{\Omega}\, \mathrm{d}v = 0, \qquad \forall\, \boldsymbol{\Omega} \in \mathrm{Skw}. \tag{156}$$

Moreover, it follows from Postulate 2, and given (101), that the external virtual power is also frame indifferent. Localizing (156), we have that $\mathbf{T} : \boldsymbol{\Omega} = 0$, implying

$$\mathbf{T} = \mathbf{T}^{\top} \quad \Leftrightarrow \quad \operatorname{ax}(\mathbf{T} - \mathbf{T}^{\top}) = \mathbf{0}. \tag{157}$$

Lastly, note that for the scalar case (100), the frame indifference postulate is automatically satisfied.[2]

2 Partwise Balances on $\mathcal{P} \subset \mathcal{B}$

We now aim to construct the partwise balances of microforces and forces. Thus, integrating (121) and (137) on a part $\mathcal{P}$ and using the divergence theorem, we arrive at

$$\int_{\mathcal{P}} (\pi^{\alpha} + \gamma^{\alpha})\, \mathrm{d}v + \int_{\partial\mathcal{P}} (\boldsymbol{\xi}^{\alpha} - \operatorname{div} \boldsymbol{\Xi}^{\alpha} + \operatorname{div}^2 \mathbb{X}^{\alpha}) \cdot \boldsymbol{n}\, \mathrm{d}a = 0, \tag{158}$$

and

$$\int_{\mathcal{P}} \boldsymbol{b}\, \mathrm{d}v + \int_{\partial\mathcal{P}} (\mathbf{T} - \operatorname{div} \mathbb{T} + \operatorname{div}^2 \mathbf{\mathbb{T}})[\boldsymbol{n}]\, \mathrm{d}a = \mathbf{0}. \tag{159}$$

Replacing the αth surface microtraction (138) and the surface traction (141) in (158) and (159), respectively, along with the surface divergence theorem for nonsmooth closed surfaces $(74)_{1,2}$, we arrive at

[1] 'Skw' is a vector space where all elements are second-order skew-symmetric tensors, that is, it alternates sign under a permutation of its vector arguments.

[2] Recall that the axial vector of a second-order skew-symmetric tensor $\frac{1}{2}(\boldsymbol{a} \otimes \boldsymbol{b} - \boldsymbol{b} \otimes \boldsymbol{a})$ is given by $\frac{1}{2}\operatorname{ax}(\boldsymbol{a} \otimes \boldsymbol{b} - \boldsymbol{b} \otimes \boldsymbol{a}) = -\frac{1}{2}\boldsymbol{a} \times \boldsymbol{b}$.

$$
\begin{aligned}
&\int\limits_{\mathcal{P}} (\pi^\alpha + \gamma^\alpha)\,\mathrm{d}v + \int\limits_{\partial\mathcal{P}} \xi_{\mathcal{S}}\,\mathrm{d}a \\
&\quad + \int\limits_{\mathcal{C}} \{\!\{\boldsymbol{\nu} \cdot ((\boldsymbol{\Xi}^\alpha - \operatorname{div}\mathbb{X}^\alpha)[\boldsymbol{n}] - \operatorname{div}_{\mathcal{S}}(\mathbb{X}^\alpha[\boldsymbol{n}]\mathbf{P}_{\!n}) + \mathbb{X}^\alpha[\boldsymbol{n}][\boldsymbol{n}] \cdot \mathbf{K}^{\top})\}\!\}\,\mathrm{d}\sigma = 0,
\end{aligned}
\tag{160}
$$

and

$$
\begin{aligned}
&\int\limits_{\mathcal{P}} \boldsymbol{b}\,\mathrm{d}v + \int\limits_{\partial\mathcal{P}} \boldsymbol{t}_{\mathcal{S}}\,\mathrm{d}a \\
&\quad + \int\limits_{\mathcal{C}} \{\!\{(\mathbb{T} - \operatorname{div}\mathbf{T})[\boldsymbol{n}][\boldsymbol{\nu}] - \operatorname{div}_{\mathcal{S}}(\mathbf{T}[\boldsymbol{n}]\mathbf{P}_{\!n})[\boldsymbol{\nu}] + \mathbf{T}[\boldsymbol{n}][\boldsymbol{n}][\mathbf{K}^{\top}[\boldsymbol{\nu}]]\}\!\}\,\mathrm{d}\sigma = \mathbf{0}.
\end{aligned}
\tag{161}
$$

Next, using the αth junction-edge microtraction (140) and the junction-edge traction (143) in (160) and (161), respectively, along with the gradient theorem for nonsmooth closed curves $(82)_{1,2}$, we arrive at

$$
\int\limits_{\mathcal{P}} (\pi^\alpha + \gamma^\alpha)\,\mathrm{d}v + \int\limits_{\partial\mathcal{P}} \xi_{\mathcal{S}}\,\mathrm{d}a + \int\limits_{\mathcal{C}} \xi^\alpha_{\mathcal{C}}\,\mathrm{d}\sigma + [\![\{\!\{\boldsymbol{\sigma} \cdot \mathbb{X}^\alpha[\boldsymbol{n}][\boldsymbol{\nu}]\}\!\}]\!]|_{\mathcal{O}} = 0, \tag{162}
$$

and

$$
\int\limits_{\mathcal{P}} \boldsymbol{b}\,\mathrm{d}v + \int\limits_{\partial\mathcal{P}} \boldsymbol{t}_{\mathcal{S}}\,\mathrm{d}a + \int\limits_{\mathcal{C}} \boldsymbol{t}_{\mathcal{C}}\,\mathrm{d}\sigma + [\![\{\!\{\mathbf{T}[\boldsymbol{n}][\boldsymbol{\nu}][\boldsymbol{\sigma}]\}\!\}]\!]|_{\mathcal{O}} = \mathbf{0}. \tag{163}
$$

Lastly, with the αth junction-point microforce (119) and the junction-point force (135), from (162) and (163), we, respectively, have that the αth partwise balance of microforces is

$$
\mathcal{F}^\flat(\mathcal{P}) := \int\limits_{\mathcal{P}} (\pi^\alpha + \gamma^\alpha)\,\mathrm{d}v + \int\limits_{\partial\mathcal{P}} \xi^\alpha_{\mathcal{S}}\,\mathrm{d}a + \int\limits_{\mathcal{C}} \xi^\alpha_{\mathcal{C}}\,\mathrm{d}\sigma + \xi^\alpha_{\mathcal{O}} = 0, \tag{164}
$$

and the partwise balance of forces is

$$
\mathcal{F}^\sharp(\mathcal{P}) := \int\limits_{\mathcal{P}} \boldsymbol{b}\,\mathrm{d}v + \int\limits_{\partial\mathcal{P}} \boldsymbol{t}_{\mathcal{S}}\,\mathrm{d}a + \int\limits_{\mathcal{C}} \boldsymbol{t}_{\mathcal{C}}\,\mathrm{d}\sigma + \boldsymbol{t}_{\mathcal{O}} = \mathbf{0}. \tag{165}
$$

Now, we aim to construct the partwise balances of microtorques and torques. To simplify computations, we use the notion of axial vectors when dealing with cross products. Multiplying (121) by $\boldsymbol{r}$ and taking the tensor product between $\boldsymbol{r}$ and (137) and then integrating these expressions on a part $\mathcal{P}$ with identities

$$\begin{aligned}\operatorname{div}(\boldsymbol{r}\otimes(\boldsymbol{\xi}^\alpha-\operatorname{div}\boldsymbol{\Xi}^\alpha+\operatorname{div}^2\mathbb{X}^\alpha)) &= \boldsymbol{\xi}^\alpha-\operatorname{div}\boldsymbol{\Xi}^\alpha+\operatorname{div}^2\mathbb{X}^\alpha\\ &\quad+\boldsymbol{r}\operatorname{div}(\boldsymbol{\xi}^\alpha-\operatorname{div}\boldsymbol{\Xi}^\alpha+\operatorname{div}^2\mathbb{X}^\alpha),\end{aligned}\tag{166}$$

and

$$\begin{aligned}\operatorname{div}(\boldsymbol{r}\otimes(\mathbf{T}-\operatorname{div}\mathbb{T}+\operatorname{div}^2\mathbf{\mathbb{T}})) &= \mathbf{T}-\operatorname{div}\mathbb{T}+\operatorname{div}^2\mathbf{\mathbb{T}}\\ &\quad+\boldsymbol{r}\otimes\operatorname{div}(\mathbf{T}-\operatorname{div}\mathbb{T}+\operatorname{div}^2\mathbf{\mathbb{T}}),\end{aligned}\tag{167}$$

followed by the application of the divergence theorem, we arrive at

$$\begin{aligned}&\int_{\mathcal{P}}(\boldsymbol{r}(\pi^\alpha+\gamma^\alpha)-\boldsymbol{\xi}^\alpha)\,\mathrm{d}v\\ &\qquad+\int_{\partial\mathcal{P}}((\boldsymbol{\Xi}^\alpha-\operatorname{div}\mathbb{X}^\alpha)[\boldsymbol{n}]+\boldsymbol{r}(\boldsymbol{\xi}^\alpha-\operatorname{div}\boldsymbol{\Xi}^\alpha+\operatorname{div}^2\mathbb{X}^\alpha)\cdot\boldsymbol{n})\,\mathrm{d}a=0,\end{aligned}\tag{168}$$

and

$$\int_{\mathcal{P}}(\boldsymbol{r}\otimes\boldsymbol{b}-\mathbf{T})\,\mathrm{d}v+\int_{\partial\mathcal{P}}((\mathbb{T}-\operatorname{div}\mathbf{\mathbb{T}})[\boldsymbol{n}]+\boldsymbol{r}\otimes(\mathbf{T}-\operatorname{div}\mathbb{T}+\operatorname{div}^2\mathbf{\mathbb{T}})[\boldsymbol{n}])\,\mathrm{d}a=\mathbf{0}.\tag{169}$$

Next, consider that

$$\operatorname{div}_{\mathcal{S}}(\boldsymbol{r}\otimes\boldsymbol{a})=\boldsymbol{r}\operatorname{div}_{\mathcal{S}}(\boldsymbol{a})+\mathbf{P}_{\!n}[\boldsymbol{a}],\tag{170}$$

yielding the following identities

$$\begin{aligned}\operatorname{div}_{\mathcal{S}}(\boldsymbol{r}\otimes\mathbf{P}_{\!n}[(\boldsymbol{\Xi}^\alpha-\operatorname{div}\mathbb{X}^\alpha)[\boldsymbol{n}]]) &= (\boldsymbol{\Xi}^\alpha-\operatorname{div}\mathbb{X}^\alpha)[\boldsymbol{n}]-(\boldsymbol{n}\otimes\boldsymbol{n})[(\boldsymbol{\Xi}^\alpha-\operatorname{div}\mathbb{X}^\alpha)[\boldsymbol{n}]]\\ &\quad+\boldsymbol{r}\operatorname{div}_{\mathcal{S}}(\mathbf{P}_{\!n}[(\boldsymbol{\Xi}^\alpha-\operatorname{div}\mathbb{X}^\alpha)[\boldsymbol{n}]]),\end{aligned}\tag{171}$$

$$\begin{aligned}\operatorname{div}_{\mathcal{S}}(\boldsymbol{r}\otimes\mathbf{P}_{\!n}[\operatorname{div}_{\mathcal{S}}(\mathbb{X}^\alpha[\boldsymbol{n}]\mathbf{P}_{\!n})]) &= \operatorname{div}_{\mathcal{S}}(\mathbb{X}^\alpha[\boldsymbol{n}]\mathbf{P}_{\!n})-(\boldsymbol{n}\otimes\boldsymbol{n})[\operatorname{div}_{\mathcal{S}}(\mathbb{X}^\alpha[\boldsymbol{n}]\mathbf{P}_{\!n})]\\ &\quad+\boldsymbol{r}\operatorname{div}_{\mathcal{S}}(\mathbf{P}_{\!n}[\operatorname{div}_{\mathcal{S}}(\mathbb{X}^\alpha[\boldsymbol{n}]\mathbf{P}_{\!n})]),\end{aligned}\tag{172}$$

and

$$\operatorname{div}_{\mathcal{S}}(\boldsymbol{r}\otimes\mathbb{X}^\alpha[\boldsymbol{n}][\boldsymbol{n}]\cdot\mathbf{K}^\top)=\mathbb{X}^\alpha[\boldsymbol{n}][\boldsymbol{n}]\cdot\mathbf{K}^\top+\boldsymbol{r}\operatorname{div}_{\mathcal{S}}(\mathbb{X}^\alpha[\boldsymbol{n}][\boldsymbol{n}]\cdot\mathbf{K}^\top).\tag{173}$$

Combining these identities with the product $\boldsymbol{r}\xi_{\mathcal{S}}^{\alpha}$, where the αth surface microtraction is given in expression (138),

$$\begin{aligned} \boldsymbol{r}\xi_{\mathcal{S}}^{\alpha} &= \boldsymbol{r}(\boldsymbol{\xi}^{\alpha} - \operatorname{div}\boldsymbol{\Xi}^{\alpha} + \operatorname{div}^2\mathbb{X}^{\alpha}) \cdot \boldsymbol{n} - \boldsymbol{r}\operatorname{div}_{\mathcal{S}}(\mathbf{P}_{n}[(\boldsymbol{\Xi}^{\alpha} - \operatorname{div}\mathbb{X}^{\alpha})[\boldsymbol{n}]]) \\ &\quad + \boldsymbol{r}\operatorname{div}_{\mathcal{S}}(\mathbf{P}_{n}[\operatorname{div}_{\mathcal{S}}(\mathbb{X}^{\alpha}[\boldsymbol{n}]\mathbf{P}_{n})]) - \boldsymbol{r}\operatorname{div}_{\mathcal{S}}(\mathbb{X}^{\alpha}[\boldsymbol{n}][\boldsymbol{n}] \cdot \mathbf{K}^{\top}), \end{aligned} \tag{174}$$

we have that

$$\begin{aligned} \boldsymbol{r}(\boldsymbol{\xi}^{\alpha} - \operatorname{div}\boldsymbol{\Xi}^{\alpha} + \operatorname{div}^2\mathbb{X}^{\alpha}) \cdot \boldsymbol{n} = \boldsymbol{r}\xi_{\mathcal{S}}^{\alpha} &+ \operatorname{div}_{\mathcal{S}}(\boldsymbol{r} \otimes \mathbf{P}_{n}[(\boldsymbol{\Xi}^{\alpha} - \operatorname{div}\mathbb{X}^{\alpha})[\boldsymbol{n}]]) \\ &- (\boldsymbol{\Xi}^{\alpha} - \operatorname{div}\mathbb{X}^{\alpha})[\boldsymbol{n}] \\ &+ (\boldsymbol{n} \otimes \boldsymbol{n})[(\boldsymbol{\Xi}^{\alpha} - \operatorname{div}\mathbb{X}^{\alpha})[\boldsymbol{n}]] \\ &- \operatorname{div}_{\mathcal{S}}(\boldsymbol{r} \otimes \mathbf{P}_{n}[\operatorname{div}_{\mathcal{S}}(\mathbb{X}^{\alpha}[\boldsymbol{n}]\mathbf{P}_{n})]) \\ &+ \operatorname{div}_{\mathcal{S}}(\mathbb{X}^{\alpha}[\boldsymbol{n}]\mathbf{P}_{n}) - (\boldsymbol{n} \otimes \boldsymbol{n})[\operatorname{div}_{\mathcal{S}}(\mathbb{X}^{\alpha}[\boldsymbol{n}]\mathbf{P}_{n})] \\ &+ \operatorname{div}_{\mathcal{S}}(\boldsymbol{r} \otimes \mathbb{X}^{\alpha}[\boldsymbol{n}][\boldsymbol{n}] \cdot \mathbf{K}^{\top}) \\ &- \mathbb{X}^{\alpha}[\boldsymbol{n}][\boldsymbol{n}] \cdot \mathbf{K}^{\top}. \end{aligned} \tag{175}$$

Similarly, consider

$$\operatorname{div}_{\mathcal{S}}(\boldsymbol{r} \otimes \mathbf{A}) = \boldsymbol{r} \otimes \operatorname{div}_{\mathcal{S}}(\mathbf{A}) + \mathbf{A}\mathbf{P}_{n}, \tag{176}$$

yielding the following identities

$$\begin{aligned} \operatorname{div}_{\mathcal{S}}(\boldsymbol{r} \otimes (\mathbb{T} - \operatorname{div}\mathbf{T})[\boldsymbol{n}]\mathbf{P}_{n}) = (\mathbb{T} - \operatorname{div}\mathbf{T})[\boldsymbol{n}] &- (\mathbb{T} - \operatorname{div}\mathbf{T})[\boldsymbol{n}](\boldsymbol{n} \otimes \boldsymbol{n}) \\ &+ \boldsymbol{r} \otimes \operatorname{div}_{\mathcal{S}}((\mathbb{T} - \operatorname{div}\mathbf{T})[\boldsymbol{n}]\mathbf{P}_{n}), \end{aligned} \tag{177}$$

$$\begin{aligned} \operatorname{div}_{\mathcal{S}}(\boldsymbol{r} \otimes (\operatorname{div}_{\mathcal{S}}(\mathbf{T}[\boldsymbol{n}]\mathbf{P}_{n})\mathbf{P}_{n}) = \operatorname{div}_{\mathcal{S}}(\mathbf{T}[\boldsymbol{n}]\mathbf{P}_{n}) &- \operatorname{div}_{\mathcal{S}}(\mathbf{T}[\boldsymbol{n}]\mathbf{P}_{n})(\boldsymbol{n} \otimes \boldsymbol{n}) \\ &+ \boldsymbol{r} \otimes \operatorname{div}_{\mathcal{S}}(\operatorname{div}_{\mathcal{S}}(\mathbf{T}[\boldsymbol{n}]\mathbf{P}_{n})\mathbf{P}_{n}), \end{aligned} \tag{178}$$

and

$$\operatorname{div}_{\mathcal{S}}(\boldsymbol{r} \otimes \mathbf{T}[\boldsymbol{n}][\boldsymbol{n}]\mathbf{K}^{\top}) = \mathbf{T}[\boldsymbol{n}][\boldsymbol{n}]\mathbf{K}^{\top} + \boldsymbol{r} \otimes \operatorname{div}_{\mathcal{S}}(\mathbf{T}[\boldsymbol{n}][\boldsymbol{n}]\mathbf{K}^{\top}). \tag{179}$$

Combining these identities with the product $\boldsymbol{r} \otimes \boldsymbol{t}_{\mathcal{S}}$, where the surface traction is given in expression (141),

$$\begin{aligned} \boldsymbol{r} \otimes \boldsymbol{t}_{\mathcal{S}} &= \boldsymbol{r} \otimes (\mathbf{T} - \operatorname{div}\mathbb{T} + \operatorname{div}^2\mathbf{T})[\boldsymbol{n}] - \boldsymbol{r} \otimes \operatorname{div}_{\mathcal{S}}((\mathbb{T} - \operatorname{div}\mathbf{T})[\boldsymbol{n}]\mathbf{P}_{n}) \\ &\quad + \boldsymbol{r} \otimes \operatorname{div}_{\mathcal{S}}(\operatorname{div}_{\mathcal{S}}(\mathbf{T}[\boldsymbol{n}]\mathbf{P}_{n})\mathbf{P}_{n}) - \boldsymbol{r} \otimes \operatorname{div}_{\mathcal{S}}(\mathbf{T}[\boldsymbol{n}][\boldsymbol{n}]\mathbf{K}^{\top}), \end{aligned} \tag{180}$$

we have that

$$
\begin{aligned}
\boldsymbol{r}\otimes(\mathbf{T}-\operatorname{div}\mathbb{T}+\operatorname{div}^2\mathbb{T})[\boldsymbol{n}] = \boldsymbol{r}\otimes\boldsymbol{t}_{\mathcal{S}} &+ \operatorname{div}_{\mathcal{S}}(\boldsymbol{r}\otimes(\mathbb{T}-\operatorname{div}\mathbb{T})[\boldsymbol{n}]\mathbf{P}_n) \\
&- (\mathbb{T}-\operatorname{div}\mathbb{T})[\boldsymbol{n}] + (\mathbb{T}-\operatorname{div}\mathbb{T})[\boldsymbol{n}](\boldsymbol{n}\otimes\boldsymbol{n}) \\
&- \operatorname{div}_{\mathcal{S}}(\boldsymbol{r}\otimes\operatorname{div}_{\mathcal{S}}(\mathbb{T}[\boldsymbol{n}]\mathbf{P}_n)\mathbf{P}_n) \\
&+ \operatorname{div}_{\mathcal{S}}(\mathbb{T}[\boldsymbol{n}]\mathbf{P}_n) - \operatorname{div}_{\mathcal{S}}(\mathbb{T}[\boldsymbol{n}]\mathbf{P}_n)(\boldsymbol{n}\otimes\boldsymbol{n}) \\
&+ \operatorname{div}_{\mathcal{S}}(\boldsymbol{r}\otimes\mathbb{T}[\boldsymbol{n}][\boldsymbol{n}]\mathbf{K}^{\top}) \\
&- \mathbb{T}[\boldsymbol{n}][\boldsymbol{n}]\mathbf{K}^{\top}.
\end{aligned} \tag{181}
$$

Using expressions (175) and (181) along with the surface divergence theorem for nonsmooth closed surfaces $(74)_{1,2}$, respectively, in (168) and (169), we arrive at

$$
\begin{aligned}
&\int_{\mathcal{P}} (\boldsymbol{r}(\pi^\alpha+\gamma^\alpha)-\boldsymbol{\xi}^\alpha)\,\mathrm{d}v \\
&+\int_{\partial\mathcal{P}} ((\boldsymbol{n}\otimes\boldsymbol{n})[(\boldsymbol{\Xi}^\alpha-\operatorname{div}\mathbb{X}^\alpha)[\boldsymbol{n}]-\operatorname{div}_{\mathcal{S}}(\mathbb{X}^\alpha[\boldsymbol{n}]\mathbf{P}_n)]-\mathbb{X}^\alpha[\boldsymbol{n}][\boldsymbol{n}]\cdot\mathbf{K}^{\top}+\boldsymbol{r}\xi^\alpha_{\mathcal{S}})\,\mathrm{d}a \\
&+\int_{\mathcal{C}} \boldsymbol{r}\{\!\!\{\boldsymbol{\nu}\cdot((\boldsymbol{\Xi}^\alpha-\operatorname{div}\mathbb{X}^\alpha)[\boldsymbol{n}]-\operatorname{div}_{\mathcal{S}}(\mathbb{X}^\alpha[\boldsymbol{n}]\mathbf{P}_n)+\mathbb{X}^\alpha[\boldsymbol{n}][\boldsymbol{n}]\cdot\mathbf{K}^{\top})\}\!\!\}\,\mathrm{d}\sigma \\
&\qquad\qquad+\int_{\mathcal{C}} \{\!\!\{\mathbb{X}^\alpha[\boldsymbol{n}][\boldsymbol{\nu}]\}\!\!\}\,\mathrm{d}\sigma = 0,
\end{aligned} \tag{182}
$$

and

$$
\begin{aligned}
&\int_{\mathcal{P}} (\boldsymbol{r}\otimes\boldsymbol{b}-\mathbf{T})\,\mathrm{d}v \\
&+\int_{\partial\mathcal{P}} (((\mathbb{T}-\operatorname{div}\mathbb{T})[\boldsymbol{n}]-\operatorname{div}_{\mathcal{S}}(\mathbb{T}[\boldsymbol{n}]\mathbf{P}_n))(\boldsymbol{n}\otimes\boldsymbol{n})-\mathbb{T}[\boldsymbol{n}][\boldsymbol{n}]\mathbf{K}^{\top}+\boldsymbol{r}\otimes\boldsymbol{t}_{\mathcal{S}})\,\mathrm{d}a \\
&+\int_{\mathcal{C}} \boldsymbol{r}\otimes\{\!\!\{(\mathbb{T}-\operatorname{div}\mathbb{T})[\boldsymbol{n}][\boldsymbol{\nu}]-\operatorname{div}_{\mathcal{S}}(\mathbb{T}[\boldsymbol{n}]\mathbf{P}_n)[\boldsymbol{\nu}]+\mathbb{T}[\boldsymbol{n}][\boldsymbol{n}][\mathbf{K}^{\top}[\boldsymbol{\nu}]]\}\!\!\}\,\mathrm{d}\sigma \\
&\qquad\qquad+\int_{\mathcal{C}} \{\!\!\{\mathbb{T}[\boldsymbol{n}][\boldsymbol{\nu}]\}\!\!\}\,\mathrm{d}\sigma = \mathbf{0}.
\end{aligned} \tag{183}
$$

Next, from identity (68), we have that $\partial_\sigma \boldsymbol{r} = \boldsymbol{\sigma}$. Thus, for any smooth scalar a and vector $\boldsymbol{a}$ fields, we have that

$$
\partial_\sigma(a\,\boldsymbol{r}) = a\,\boldsymbol{\sigma} + \boldsymbol{r}\,\partial_\sigma a, \tag{184}
$$

and

$$\partial_\sigma(\boldsymbol{r} \otimes \boldsymbol{a}) = \boldsymbol{\sigma} \otimes \boldsymbol{a} + \boldsymbol{r} \otimes \partial_\sigma \boldsymbol{a}. \tag{185}$$

Replacing the αth junction-edge microtraction (140) and the junction-edge traction (143) in expression (182) and (183) while considering the identities (184) and (185), respectively, in the following forms

$$\partial_\sigma(\boldsymbol{r}(\boldsymbol{\sigma} \cdot \mathbb{X}^\alpha[\boldsymbol{n}][\boldsymbol{\nu}])) = (\boldsymbol{\sigma} \cdot \mathbb{X}^\alpha[\boldsymbol{n}][\boldsymbol{\nu}])\boldsymbol{\sigma} + \boldsymbol{r}\, \partial_\sigma(\boldsymbol{\sigma} \cdot \mathbb{X}^\alpha[\boldsymbol{n}][\boldsymbol{\nu}]), \tag{186}$$

and

$$\partial_\sigma(\boldsymbol{r} \otimes \mathbb{T}[\boldsymbol{n}][\boldsymbol{\nu}][\boldsymbol{\sigma}]) = \boldsymbol{\sigma} \otimes \mathbb{T}[\boldsymbol{n}][\boldsymbol{\nu}][\boldsymbol{\sigma}] + \boldsymbol{r} \otimes \partial_\sigma(\mathbb{T}[\boldsymbol{n}][\boldsymbol{\nu}][\boldsymbol{\sigma}]), \tag{187}$$

and using the gradient theorem on nonsmooth closed curves $(82)_{1,2}$, we are led to

$$\begin{aligned}
&\int_{\mathcal{P}} (\boldsymbol{r}(\pi^\alpha + \gamma^\alpha) - \boldsymbol{\xi}^\alpha)\, \mathrm{d}v \\
&+ \int_{\partial\mathcal{P}} ((\boldsymbol{n} \otimes \boldsymbol{n})[(\Xi^\alpha - \operatorname{div} \mathbb{X}^\alpha)[\boldsymbol{n}] - \operatorname{div}_{\mathcal{S}}(\mathbb{X}^\alpha[\boldsymbol{n}]\mathbf{P}_n)] - \mathbb{X}^\alpha[\boldsymbol{n}][\boldsymbol{n}] \cdot \mathbf{K}^\top + \boldsymbol{r}\xi^\alpha_{\mathcal{S}})\, \mathrm{d}a \\
&+ \int_{\mathcal{C}} \{\!\!\{\mathbb{X}^\alpha[\boldsymbol{n}][\boldsymbol{\nu}] - (\boldsymbol{\sigma} \cdot \mathbb{X}^\alpha[\boldsymbol{n}][\boldsymbol{\nu}])\boldsymbol{\sigma}\}\!\!\}\mathrm{d}\sigma \\
&+ \int_{\mathcal{C}} \boldsymbol{r}\xi_{\mathcal{C}}\, \mathrm{d}\sigma + \boldsymbol{r}_{\circ}[\![\{\!\!\{\boldsymbol{\sigma} \cdot \mathbb{X}^\alpha[\boldsymbol{n}][\boldsymbol{\nu}]\}\!\!\}]\!]|_{\circ} = 0,
\end{aligned} \tag{188}$$

and

$$\begin{aligned}
&\int_{\mathcal{P}} (\boldsymbol{r} \otimes \boldsymbol{b} - \mathbf{T})\, \mathrm{d}v \\
&+ \int_{\partial\mathcal{P}} (((\mathbb{T} - \operatorname{div} \mathbb{T})[\boldsymbol{n}] - \operatorname{div}_{\mathcal{S}}(\mathbb{T}[\boldsymbol{n}]\mathbf{P}_n))(\boldsymbol{n} \otimes \boldsymbol{n}) - \mathbb{T}[\boldsymbol{n}][\boldsymbol{n}]\mathbf{K}^\top + \boldsymbol{r} \otimes \boldsymbol{t}_{\mathcal{S}})\, \mathrm{d}a \\
&+ \int_{\mathcal{C}} \{\!\!\{\mathbb{T}[\boldsymbol{n}][\boldsymbol{\nu}] - \boldsymbol{\sigma} \otimes \mathbb{T}[\boldsymbol{n}][\boldsymbol{\nu}][\boldsymbol{\sigma}]\}\!\!\}\mathrm{d}\sigma \\
&+ \int_{\mathcal{C}} \boldsymbol{r} \otimes \boldsymbol{t}_{\mathcal{C}}\, \mathrm{d}\sigma + \boldsymbol{r}_{\circ} \otimes [\![\{\!\!\{\mathbb{T}[\boldsymbol{n}][\boldsymbol{\nu}][\boldsymbol{\sigma}]\}\!\!\}]\!]_{\circ} = \mathbf{0}.
\end{aligned} \tag{189}$$

Moreover, expression (189) may be multiplied by negative one and summed to its transpose to arrive at

$$
\begin{aligned}
&\int_{\mathcal{P}} (\boldsymbol{r} \times \boldsymbol{b} + \operatorname{ax}(\mathbf{T} - \mathbf{T}^{\top}))\, \mathrm{d}v \\
&\qquad + \int_{\partial\mathcal{P}} \boldsymbol{n} \times ((\mathbb{T} - \operatorname{div}\mathbf{\mathbb{T}})[\boldsymbol{n}][\boldsymbol{n}] - \operatorname{div}_{\mathcal{S}}(\mathbf{\mathbb{T}}[\boldsymbol{n}]\mathbf{P}_{\!n})[\boldsymbol{n}])\, \mathrm{d}a \\
&\qquad\quad + \int_{\partial\mathcal{P}} (-2 \operatorname{ax}(\operatorname{skw}(\mathbf{\mathbb{T}}[\boldsymbol{n}][\boldsymbol{n}]\mathbf{K}^{\top})^{\top}) + \boldsymbol{r} \times \boldsymbol{t}_{\mathcal{S}})\, \mathrm{d}a \\
&\qquad + \int_{\mathcal{C}} 2 \operatorname{ax}(\operatorname{skw}(\{\!\!\{\mathbf{\mathbb{T}}[\boldsymbol{n}][\boldsymbol{\nu}] - \boldsymbol{\sigma} \otimes \mathbf{\mathbb{T}}[\boldsymbol{n}][\boldsymbol{\nu}][\boldsymbol{\sigma}]\}\!\!\})^{\top})\, \mathrm{d}\sigma \\
&\qquad\qquad + \int_{\mathcal{C}} \boldsymbol{r} \times \boldsymbol{t}_{\mathcal{C}}\, \mathrm{d}\sigma + \boldsymbol{r}_{\mathcal{O}} \times [\![\{\!\!\{\mathbf{\mathbb{T}}[\boldsymbol{n}][\boldsymbol{\nu}][\boldsymbol{\sigma}]\}\!\!\}]\!]|_{\mathcal{O}} = \mathbf{0}. \qquad (190)
\end{aligned}
$$

Additionally, the terms multiplied by $\boldsymbol{r}_{\mathcal{O}}$, in expressions (188) and (190), may be identified as the αth junction-point microforce (119) and the junction-point force (135), respectively.

We are now in a position to stipulate the explicit form of the couple-fields as follows. The αth surface-couple microtraction $\varpi^{\alpha}_{\mathcal{S}} := \varpi^{\alpha}_{\mathcal{S}}(\boldsymbol{x}, t; \boldsymbol{n}, \mathbf{K})$ and the αth junction-edge-couple traction $\varpi^{\alpha}_{\mathcal{C}} := \varpi^{\alpha}_{\mathcal{C}}(\boldsymbol{x}, t; \boldsymbol{n}, \boldsymbol{\nu})$ are given by

$$
\begin{cases}
\varpi^{\alpha}_{\mathcal{S}}(\boldsymbol{x}, t; \boldsymbol{n}, \mathbf{K}) := (\boldsymbol{n} \otimes \boldsymbol{n})[(\Xi^{\alpha} - \operatorname{div}\mathbb{X}^{\alpha})[\boldsymbol{n}] - \operatorname{div}_{\mathcal{S}}(\mathbb{X}^{\alpha}[\boldsymbol{n}]\mathbf{P}_{\!n})] \\
\qquad\qquad - \mathbb{X}^{\alpha}[\boldsymbol{n}][\boldsymbol{n}] \cdot \mathbf{K}^{\top}, \\
\varpi^{\alpha}_{\mathcal{C}}(\boldsymbol{x}, t; \boldsymbol{n}^{\pm}) := \{\!\!\{\mathbf{P}_{\!\sigma}[\mathbb{X}^{\alpha}[\boldsymbol{n}][\boldsymbol{\nu}]]\}\!\!\},
\end{cases}
\qquad (191)
$$

where $\mathbf{P}_{\!\sigma} := \mathbf{1} - \boldsymbol{\sigma} \otimes \boldsymbol{\sigma}$.

Analogously, the surface-couple traction $\boldsymbol{m}_{\mathcal{S}} := \boldsymbol{m}_{\mathcal{S}}(\boldsymbol{x}, t; \boldsymbol{n}, \mathbf{K})$ and the junction-edge-couple traction $\boldsymbol{m}_{\mathcal{C}} := \boldsymbol{m}_{\mathcal{C}}(\boldsymbol{x}, t; \boldsymbol{n}, \boldsymbol{\nu})$ are given by

$$
\begin{cases}
\boldsymbol{m}_{\mathcal{S}}(\boldsymbol{x}, t; \boldsymbol{n}, \mathbf{K}) := \boldsymbol{n} \times ((\mathbb{T} - \operatorname{div}\mathbf{\mathbb{T}})[\boldsymbol{n}][\boldsymbol{n}] - \operatorname{div}_{\mathcal{S}}(\mathbf{\mathbb{T}}[\boldsymbol{n}]\mathbf{P}_{\!n})[\boldsymbol{n}]) \\
\qquad\qquad - 2 \operatorname{ax}(\operatorname{skw}(\mathbf{\mathbb{T}}[\boldsymbol{n}][\boldsymbol{n}]\mathbf{K}^{\top})^{\top}), \\
\boldsymbol{m}_{\mathcal{C}}(\boldsymbol{x}, t; \boldsymbol{n}^{\pm}) := 2 \operatorname{ax}(\operatorname{skw}(\{\!\!\{\mathbf{\mathbb{T}}[\boldsymbol{n}][\boldsymbol{\nu}]\}\!\!\})^{\top}) - \boldsymbol{\sigma} \times \{\!\!\{\mathbf{\mathbb{T}}[\boldsymbol{n}][\boldsymbol{\nu}][\boldsymbol{\sigma}]\}\!\!\}.
\end{cases}
\qquad (192)
$$

Moreover, we have that $\operatorname{ax}(\mathbf{T} - \mathbf{T}^{\top}) = \mathbf{0}$ from frame indifference (157), where this equation then represents the pointwise balance of torques.

Thus, on a part $\mathcal{P} \subset \mathcal{B}$, we have that the αth volume balance of microtorques reads

$$\mathcal{T}^{\flat}(\mathcal{P}) := \int_{\mathcal{P}} (\boldsymbol{r}(\pi^{\alpha} + \gamma^{\alpha}) - \boldsymbol{\xi}^{\alpha})\,\mathrm{d}v + \int_{\partial\mathcal{P}} \boldsymbol{\varpi}_{\mathcal{S}}^{\alpha}\,\mathrm{d}a + \int_{\partial\mathcal{P}} \boldsymbol{r}\xi_{\mathcal{S}}^{\alpha}\,\mathrm{d}a$$
$$+ \int_{\mathcal{C}} \boldsymbol{\varpi}_{\mathcal{C}}^{\alpha}\,\mathrm{d}\sigma + \int_{\mathcal{C}} \boldsymbol{r}\xi_{\mathcal{C}}^{\alpha}\,\mathrm{d}\sigma + \boldsymbol{r}_{\mathcal{O}}\xi_{\mathcal{O}}^{\alpha} = \mathbf{0}, \tag{193}$$

while the partwise balance of torques reads

$$\mathcal{T}^{\sharp}(\mathcal{P}) := \int_{\mathcal{P}} \boldsymbol{r} \times \boldsymbol{b}\,\mathrm{d}v + \int_{\partial\mathcal{P}} \boldsymbol{m}_{\mathcal{S}}\,\mathrm{d}a + \int_{\partial\mathcal{P}} \boldsymbol{r} \times \boldsymbol{t}_{\mathcal{S}}\,\mathrm{d}a$$
$$+ \int_{\mathcal{C}} \boldsymbol{m}_{\mathcal{C}}\,\mathrm{d}\sigma + \int_{\mathcal{C}} \boldsymbol{r} \times \boldsymbol{t}_{\mathcal{C}}\,\mathrm{d}\sigma + \boldsymbol{r}_{\mathcal{O}} \times \boldsymbol{t}_{\mathcal{O}} = \mathbf{0}. \tag{194}$$

3 Action-Reaction Principle

An important consequence of the representations of the surface microtractions, the surface traction, the surface-couple microtractions, and the surface-couple traction is that they are local at any point $\boldsymbol{x}$ on $\mathcal{S}$ and any time t. Analogously, the junction-edge microtractions, the junction-edge traction, the junction-edge-couple microtractions, and the junction-edge-couple traction are local at any point $\boldsymbol{x}$ on $\mathcal{C}$ and any time t. Previously, we stated that $\xi_{\mathcal{S}}$ and $\boldsymbol{t}_{\mathcal{S}}$ depend on $\mathcal{S}$ through the unit normal $\boldsymbol{n}$, the curvature tensor $\mathbf{K}$, and the second curvature tensor $\mathbb{K}$ of $\mathcal{S}$ at $\boldsymbol{x}$, whereas $\boldsymbol{\varpi}_{\mathcal{S}}$ and $\boldsymbol{m}_{\mathcal{S}}$ depend on $\mathcal{S}$ only through the unit normal $\boldsymbol{n}$ and the curvature tensor $\mathbf{K}$ of $\mathcal{S}$ at $\boldsymbol{x}$.[3] Conversely, $\xi_{\mathcal{C}}$ and $\boldsymbol{t}_{\mathcal{C}}$ depend on $\mathcal{C}$ through the unit normals $\{\boldsymbol{n}^{+}, \boldsymbol{n}^{-}\}$ (or equivalently through the unit tangent-normals $\{\boldsymbol{\nu}^{+}, \boldsymbol{\nu}^{-}\}$) and the curvature tensors $\{\mathbf{K}^{+}, \mathbf{K}^{-}\}$, whereas $\boldsymbol{\varpi}_{\mathcal{C}}$ and $\boldsymbol{m}_{\mathcal{C}}$ depend on $\mathcal{C}$ only through the unit normals $\{\boldsymbol{n}^{+}, \boldsymbol{n}^{-}\}$.

Next, let $-\mathcal{S}$ denote the surface $\mathcal{S}$ and $-\mathcal{C}$ the junction-edge $\mathcal{C}$ oriented, respectively, by $-\boldsymbol{n}$ and $\{-\boldsymbol{n}^{+}, -\boldsymbol{n}^{-}\}$.[4] With reference to (36) and (37), $-\mathcal{S}$ has a curvature tensor $-\mathbf{K}$ and a second curvature tensor $-\mathbb{K}$, and we see that

$$\begin{cases} \xi_{\mathcal{S}}(\boldsymbol{x}, t; -\boldsymbol{n}, -\mathbf{K}, -\mathbb{K}) = -\xi_{\mathcal{S}}(\boldsymbol{x}, t; \boldsymbol{n}, \mathbf{K}, \mathbb{K}), \\ \boldsymbol{t}_{\mathcal{S}}(\boldsymbol{x}, t; -\boldsymbol{n}, -\mathbf{K}, -\mathbb{K}) = -\boldsymbol{t}_{\mathcal{S}}(\boldsymbol{x}, t; \boldsymbol{n}, \mathbf{K}, \mathbb{K}), \\ \boldsymbol{\varpi}_{\mathcal{S}}(\boldsymbol{x}, t; -\boldsymbol{n}, -\mathbf{K}) = -\boldsymbol{\varpi}_{\mathcal{S}}(\boldsymbol{x}, t; \boldsymbol{n}, \mathbf{K}), \\ \boldsymbol{m}_{\mathcal{S}}(\boldsymbol{x}, t; -\boldsymbol{n}, -\mathbf{K}) = -\boldsymbol{m}_{\mathcal{S}}(\boldsymbol{x}, t; \boldsymbol{n}, \mathbf{K}), \end{cases} \tag{195}$$

[3] In Toupin's theory [15, 16] as well as in Fried & Gurtin's theory, the junction-edge traction depends only on the normal and the junction-edge-couple traction is not present.

[4] The change from $\boldsymbol{n}^{\pm}$ to $-\boldsymbol{n}^{\pm}$ on $\mathcal{C}$ also implies the change from $\boldsymbol{\sigma}$ to $-\boldsymbol{\sigma}$. Similarly, with reference to $(64)_1$ and $(65)_1$, the tangential partial derivative ∂_{σ} also changes its sign.

and

$$\begin{cases} \xi_{\mathcal{C}}(\boldsymbol{x}, t; -\boldsymbol{n}^+, -\boldsymbol{n}^-, -\mathbf{K}^+, -\mathbf{K}^-) = -\xi_{\mathcal{C}}(\boldsymbol{x}, t; \boldsymbol{n}^+, \boldsymbol{n}^-, \mathbf{K}^+, \mathbf{K}^-), \\ \boldsymbol{t}_{\mathcal{C}}(\boldsymbol{x}, t; -\boldsymbol{n}^+, -\boldsymbol{n}^-, -\mathbf{K}^+, -\mathbf{K}^-) = -\boldsymbol{t}_{\mathcal{C}}(\boldsymbol{x}, t; \boldsymbol{n}^+, \boldsymbol{n}^-, \mathbf{K}^+, \mathbf{K}^-), \\ \varpi_{\mathcal{C}}(\boldsymbol{x}, t; -\boldsymbol{n}^+, -\boldsymbol{n}^-) = -\varpi_{\mathcal{C}}(\boldsymbol{x}, t; \boldsymbol{n}^+, \boldsymbol{n}^-), \\ \boldsymbol{m}_{\mathcal{C}}(\boldsymbol{x}, t; -\boldsymbol{n}^+, -\boldsymbol{n}^-) = -\boldsymbol{m}_{\mathcal{C}}(\boldsymbol{x}, t; \boldsymbol{n}^+, \boldsymbol{n}^-). \end{cases} \tag{196}$$

The relations (195) make explicit the action-reaction principle in terms of surface and surface-couple, microtractions and tractions, between two smooth surfaces endowed with opposite unit normals, opposite curvature tensors, and opposite second curvature tensors at a point. Conversely, relations (196) present the action-reaction principle in terms of junction-edge and junction-edge-couple, microtractions and tractions, between two parts of the same oriented nonsmooth surface divided by a junction-edge, which in turn is defined by a pair of limiting unit normals at a point.

4 Balances on $\mathcal{S} \subset \mathcal{P}$

In dealing with an open surface $\mathcal{S}$, additional interactions are developed and represented by the αth boundary-edge microtraction $\xi^{\alpha}_{\partial\mathcal{S}}$, the boundary-edge traction $\boldsymbol{t}_{\partial\mathcal{S}}$, the αth boundary-edge couple-microtraction $\varpi^{\alpha}_{\partial\mathcal{S}}$, and the boundary-edge couple-traction $\boldsymbol{m}_{\partial\mathcal{S}}$ on $\partial\mathcal{S}$. These quantities may be obtained by specializing the set $\{\xi^{\alpha}_{\mathcal{C}}, \boldsymbol{t}_{\mathcal{C}}, \varpi^{\alpha}_{\mathcal{C}}, \boldsymbol{m}_{\mathcal{C}}\}$ of microtractions, tractions, couple-microtractions, and couple-tractions developed on a junction-edge $\mathcal{C}$ if one splits the surface by its junction-edge. Thus, these fields read

$$\begin{aligned} \xi^{\alpha}_{\partial\mathcal{S}}(\boldsymbol{x}, t; \boldsymbol{n}, \boldsymbol{\nu}, \mathbf{K}) := \boldsymbol{\nu} \cdot ((\boldsymbol{\Xi}^{\alpha} - \operatorname{div} \mathbb{X}^{\alpha})[\boldsymbol{n}] - \operatorname{div}_{\mathcal{S}}(\mathbb{X}^{\alpha}[\boldsymbol{n}]\mathbf{P}_{\!n}) \\ + \mathbb{X}^{\alpha}[\boldsymbol{n}][\boldsymbol{n}] \cdot \mathbf{K}^{\top}) - \partial_{\sigma}(\boldsymbol{\sigma} \cdot \mathbb{X}^{\alpha}[\boldsymbol{n}][\boldsymbol{\nu}]), \end{aligned} \tag{197}$$

$$\begin{aligned} \boldsymbol{t}_{\partial\mathcal{S}}(\boldsymbol{x}, t; \boldsymbol{n}, \boldsymbol{\nu}, \mathbf{K}) := (\mathbb{T} - \operatorname{div} \mathbf{T})[\boldsymbol{n}][\boldsymbol{\nu}] - \operatorname{div}_{\mathcal{S}}(\mathbf{T}[\boldsymbol{n}]\mathbf{P}_{\!n})[\boldsymbol{\nu}] \\ + \mathbf{T}[\boldsymbol{n}][\boldsymbol{n}][\mathbf{K}^{\top}[\boldsymbol{\nu}]] - \partial_{\sigma}(\mathbf{T}[\boldsymbol{n}][\boldsymbol{\nu}][\boldsymbol{\sigma}]), \end{aligned} \tag{198}$$

$$\varpi^{\alpha}_{\partial\mathcal{S}}(\boldsymbol{x}, t; \boldsymbol{n}, \boldsymbol{\nu}) := \mathbf{P}_{\!\sigma}[\mathbb{X}^{\alpha}[\boldsymbol{n}][\boldsymbol{\nu}]], \tag{199}$$

and

$$\boldsymbol{m}_{\partial\mathcal{S}}(\boldsymbol{x}, t; \boldsymbol{n}, \boldsymbol{\nu}) := 2 \operatorname{ax}(\operatorname{skw}(\mathbf{T}[\boldsymbol{n}][\boldsymbol{\nu}])^{\top}) - \boldsymbol{\sigma} \times \mathbf{T}[\boldsymbol{n}][\boldsymbol{\nu}][\boldsymbol{\sigma}]. \tag{200}$$

Furthermore, additional interactions may be developed at the boundary- and junction-points along $\partial\mathcal{S}$. At these boundary- and junction-points $\partial^2\mathcal{S} \equiv \partial\mathcal{C} \in \partial\mathcal{S}$, the αth junction-point microforce $\xi_{\partial^2\mathcal{S}}$, the junction-point force $\boldsymbol{t}_{\partial^2\mathcal{S}}$, the αth boundary-point microforce $\xi^{\alpha}_{\partial\mathcal{C}}$, and the boundary-point force $\boldsymbol{t}_{\partial\mathcal{C}}$ read

$$\xi^{\alpha}_{\partial^2\mathcal{S}}(\boldsymbol{x}, t; {}^{\pm}\boldsymbol{n}) := [\![\boldsymbol{\sigma} \cdot \mathbb{X}^{\alpha}[\boldsymbol{n}][\boldsymbol{\nu}]]\!], \tag{201}$$

$$\boldsymbol{t}_{\partial^2\mathcal{S}}(\boldsymbol{x}, t; {}^{\pm}\boldsymbol{n}) := [\![\mathbb{T}[\boldsymbol{n}][\boldsymbol{\nu}][\boldsymbol{\sigma}]]\!], \tag{202}$$

$$\xi^{\alpha}_{\partial\mathcal{C}}(\boldsymbol{x}, t; \boldsymbol{n}^{\pm}) :=]\!\{\boldsymbol{\sigma} \cdot \mathbb{X}^{\alpha}[\boldsymbol{n}][\boldsymbol{\nu}]\}\![, \tag{203}$$

and

$$\boldsymbol{t}_{\partial\mathcal{C}}(\boldsymbol{x}, t; \boldsymbol{n}^{\pm}) :=]\!\{\mathbb{T}[\boldsymbol{n}][\boldsymbol{\nu}][\boldsymbol{\sigma}]\}\![. \tag{204}$$

With these additional fields, we postulate the balance of microforces and forces as well as the balance of microtorques and torques on nonsmooth open surfaces as follows.

Postulate 3 *(*αth partwise surface balance of microforces and partwise surface balance of forces*) On a nonsmooth open oriented surface $\mathcal{S} \subset \mathcal{B}$, we postulate that the αth partwise surface balance of microforces*

$$\mathcal{F}^{\flat}(\mathcal{S}) := \int\limits_{\mathcal{S}} (\xi^{\alpha}_{\mathcal{S}} + \xi^{\alpha}_{\mathcal{S}^*})\,\mathrm{d}a + \int\limits_{\partial\mathcal{S}} \xi^{\alpha}_{\partial\mathcal{S}}\,\mathrm{d}\sigma + \int\limits_{\mathcal{C}} \xi^{\alpha}_{\mathcal{C}}\,\mathrm{d}\sigma + \xi^{\alpha}_{\partial^2\mathcal{S}} + \xi^{\alpha}_{\partial\mathcal{C}} + \xi^{\alpha}_{\mathcal{O}} \stackrel{!}{=} 0, \tag{205}$$

$$\forall\, \mathcal{S} \subset \mathcal{B} \quad \textit{and} \quad \forall\, t, \quad \textit{with} \quad 1 \le \alpha \le n,$$

and the partwise surface balance of forces

$$\mathcal{F}^{\sharp}(\mathcal{S}) := \int\limits_{\mathcal{S}} (\boldsymbol{t}_{\mathcal{S}} + \boldsymbol{t}_{\mathcal{S}^*})\,\mathrm{d}a + \int\limits_{\partial\mathcal{S}} \boldsymbol{t}_{\partial\mathcal{S}}\,\mathrm{d}\sigma + \int\limits_{\mathcal{C}} \boldsymbol{t}_{\mathcal{C}}\,\mathrm{d}\sigma + \boldsymbol{t}_{\partial^2\mathcal{S}} + \boldsymbol{t}_{\partial\mathcal{C}} + \boldsymbol{t}_{\mathcal{O}} \stackrel{!}{=} \boldsymbol{0}, \tag{206}$$

$$\forall\, \mathcal{S} \subset \mathcal{B} \quad \textit{and} \quad \forall\, t,$$

hold. Here, $\xi^{\alpha}_{\mathcal{S}^*}$ and $\boldsymbol{t}_{\mathcal{S}^*}$ are, respectively, the αth surface microtraction and surface traction developed on the opposite side of $\mathcal{S}$, that is, $\mathcal{S}^*$.

Postulate 4 *(*αth partwise surface balance of microtorques and partwise surface balance of torques*) On a smooth open oriented surface $\mathcal{S} \subset \mathcal{B}$, we postulate that the αth partwise surface balance of microtorques*

$$
\begin{aligned}
\mathcal{T}^{\flat}(\mathcal{S}) := & \int_{\mathcal{S}} (\varpi^{\alpha}_{\mathcal{S}} + \varpi^{\alpha}_{\mathcal{S}^*}) \,\mathrm{d}a + \int_{\mathcal{S}} \boldsymbol{r}(\xi^{\alpha}_{\mathcal{S}} + \xi^{\alpha}_{\mathcal{S}^*}) \,\mathrm{d}a \\
& + \int_{\partial\mathcal{S}} \varpi^{\alpha}_{\partial\mathcal{S}} \,\mathrm{d}\sigma + \int_{\partial\mathcal{S}} \boldsymbol{r}\xi^{\alpha}_{\partial\mathcal{S}} \,\mathrm{d}\sigma + \int_{\mathcal{C}} \varpi^{\alpha}_{\mathcal{C}} \,\mathrm{d}\sigma + \int_{\mathcal{C}} \boldsymbol{r}\xi^{\alpha}_{\mathcal{C}} \,\mathrm{d}\sigma \\
& + \boldsymbol{r}_{\partial^2\mathcal{S}}\xi^{\alpha}_{\partial^2\mathcal{S}} + \boldsymbol{r}_{\partial\mathcal{C}}\xi^{\alpha}_{\partial\mathcal{C}} + \boldsymbol{r}_{\mathcal{O}}\xi^{\alpha}_{\mathcal{O}} \stackrel{!}{=} \mathbf{0}, \qquad (207)
\end{aligned}
$$

$$
\forall \mathcal{S} \subset \mathcal{B} \quad \textit{and} \quad \forall t, \quad \textit{with} \quad 1 \leq \alpha \leq n,
$$

and the partwise surface balance of torques

$$
\begin{aligned}
\mathcal{T}^{\sharp}(\mathcal{S}) := & \int_{\mathcal{S}} (\boldsymbol{m}_{\mathcal{S}} + \boldsymbol{m}_{\mathcal{S}^*}) \,\mathrm{d}a + \int_{\mathcal{S}} \boldsymbol{r} \times (\boldsymbol{t}_{\mathcal{S}} + \boldsymbol{t}_{\mathcal{S}^*}) \,\mathrm{d}a \\
& + \int_{\partial\mathcal{S}} \boldsymbol{m}_{\partial\mathcal{S}} \,\mathrm{d}\sigma + \int_{\partial\mathcal{S}} \boldsymbol{r} \times \boldsymbol{t}_{\partial\mathcal{S}} \,\mathrm{d}\sigma + \int_{\mathcal{C}} \boldsymbol{m}_{\mathcal{C}} \,\mathrm{d}\sigma + \int_{\mathcal{C}} \boldsymbol{r} \times \boldsymbol{t}_{\mathcal{C}} \,\mathrm{d}\sigma \\
& + \boldsymbol{r}_{\partial^2\mathcal{S}} \times \boldsymbol{t}_{\partial^2\mathcal{S}} + \boldsymbol{r}_{\partial\mathcal{C}} \times \boldsymbol{t}_{\partial\mathcal{C}} + \boldsymbol{r}_{\mathcal{O}} \times \boldsymbol{t}_{\mathcal{O}} \stackrel{!}{=} \mathbf{0}, \qquad (208)
\end{aligned}
$$

$$
\forall \mathcal{S} \subset \mathcal{B} \quad \textit{and} \quad \forall t,
$$

hold. Here, $\varpi^{\alpha}_{\mathcal{S}^*}$ and $\boldsymbol{m}_{\mathcal{S}^*}$ are, respectively, the αth surface-couple microtraction and surface-couple traction developed on the opposite side of $\mathcal{S}$, that is, $\mathcal{S}^*$.

5 Jump Conditions

5.1 Jump Conditions Arising from Surface Balances

Consider the αth partwise surface balance of microforces (205) and the partwise surface balance of forces (206). Replacing in these surface balances the αth boundary-point microforce (203), the boundary-point force (204), the αth junction-point microforces (119) and (201), the junction-point forces (135) and (202), the αth boundary-edge microtraction (197), the boundary-edge traction (198), the αth junction-edge microtraction (140), the junction-edge traction (143), and using the gradient theorem on nonsmooth open curves $(83)_{1,2}$ along with the surface divergence theorem on nonsmooth open surfaces $(77)_{1,2}$, we have that

$$\int_{\mathcal{S}} (\xi^{\alpha}_{\mathcal{S}} + \xi^{\alpha}_{\mathcal{S}*} + \mathrm{div}_{\mathcal{S}}(\mathbf{P}_{n}[(\Xi^{\alpha} - \mathrm{div}\,\mathbb{X}^{\alpha})[\boldsymbol{n}]])$$

$$- \mathrm{div}_{\mathcal{S}}(\mathbf{P}_{n}[\mathrm{div}_{\mathcal{S}}(\mathbb{X}^{\alpha}[\boldsymbol{n}]\mathbf{P}_{n})]) + \mathrm{div}_{\mathcal{S}}(\mathbb{X}^{\alpha}[\boldsymbol{n}][\boldsymbol{n}] \cdot \mathbf{K}^{\top}))\, \mathrm{d}a = 0, \tag{209}$$

and

$$\int_{\mathcal{S}} (\boldsymbol{t}_{\mathcal{S}} + \boldsymbol{t}_{\mathcal{S}^*} + \mathrm{div}_{\mathcal{S}}((\mathbb{T} - \mathrm{div}\,\mathbf{T})[\boldsymbol{n}]\mathbf{P}_{n})$$

$$- \mathrm{div}_{\mathcal{S}}(\mathrm{div}_{\mathcal{S}}(\mathbf{T}[\boldsymbol{n}]\mathbf{P}_{n})\mathbf{P}_{n}) + \mathrm{div}_{\mathcal{S}}(\mathbf{T}[\boldsymbol{n}][\boldsymbol{n}]\mathbf{K}^{\top}))\, \mathrm{d}a = \mathbf{0}, \tag{210}$$

and by localizing these expressions, we arrive at the following jump conditions across an open surface $\mathcal{S}$,

$$\xi^{\alpha}_{\mathcal{S}} + \xi^{\alpha}_{\mathcal{S}*} = -\mathrm{div}_{\mathcal{S}}(\mathbf{P}_{n}[(\Xi^{\alpha} + \mathrm{div}\,\mathbb{X}^{\alpha})[\boldsymbol{n}]])$$

$$+ \mathrm{div}_{\mathcal{S}}(\mathbf{P}_{n}[\mathrm{div}_{\mathcal{S}}(\mathbb{X}^{\alpha}[\boldsymbol{n}]\mathbf{P}_{n})]) - \mathrm{div}_{\mathcal{S}}(\mathbb{X}^{\alpha}[\boldsymbol{n}][\boldsymbol{n}] \cdot \mathbf{K}^{\top}), \tag{211}$$

and

$$\boldsymbol{t}_{\mathcal{S}} + \boldsymbol{t}_{\mathcal{S}^*} = -\mathrm{div}_{\mathcal{S}}((\mathbb{T} - \mathrm{div}\,\mathbf{T})[\boldsymbol{n}]\mathbf{P}_{n})$$

$$+ \mathrm{div}_{\mathcal{S}}(\mathrm{div}_{\mathcal{S}}(\mathbf{T}[\boldsymbol{n}]\mathbf{P}_{n})\mathbf{P}_{n}) - \mathrm{div}_{\mathcal{S}}(\mathbf{T}[\boldsymbol{n}][\boldsymbol{n}]\mathbf{K}^{\top}). \tag{212}$$

Moreover, with the expressions for the αth surface microtraction (138) and the surface traction (141), we have that

$$\xi^{\alpha}_{\mathcal{S}*}(\boldsymbol{x}, t; \boldsymbol{n}) = -(\boldsymbol{\xi}^{\alpha} - \mathrm{div}\,\Xi^{\alpha} + \mathrm{div}^{2}\mathbb{X}^{\alpha}) \cdot \boldsymbol{n}, \tag{213}$$

and

$$\boldsymbol{t}_{\mathcal{S}^*}(\boldsymbol{x}, t; \boldsymbol{n}) = -(\mathbf{T} - \mathrm{div}\,\mathbb{T} + \mathrm{div}^{2}\mathbf{T})[\boldsymbol{n}]. \tag{214}$$

Next, consider the αth partwise surface balance of microtorques (207) and the partwise surface balance of torques (208). Replacing in these surface balances the αth boundary-point microforce (203), the boundary-point force (204), the αth junction-point microforces (119) and (201), the junction-point forces (135) and (202), the αth boundary-edge microtraction (197), the boundary-edge traction (198), the αth junction-edge microtraction (140), the junction-edge traction (143), the αth boundary-edge-couple microtraction (199), the boundary-edge-couple traction (200), the αth junction-edge-couple microtraction $(191)_2$, the junction-edge-couple traction $(192)_2$, the jump conditions (211) and (212), with identity (170) and identity (176) in the form

$$\mathrm{div}_{\mathcal{S}}(\boldsymbol{r} \times \mathbf{A}) = \boldsymbol{r} \times \mathrm{div}_{\mathcal{S}}(\mathbf{A}) + 2\,\mathrm{ax}\,(\mathrm{skw}\,(\mathbf{A}\mathbf{P}_{\!n})^{\top}), \tag{215}$$

and using the gradient theorem on nonsmooth open curves $(83)_{1,2}$ along with the surface divergence theorem on nonsmooth open surfaces $(77)_{1,2}$, we have that

$$\begin{aligned}\int\limits_{\mathcal{S}} (\varpi^{\alpha}_{\mathcal{S}} + \varpi^{\alpha}_{\mathcal{S}^*} &+ \mathrm{div}_{\mathcal{S}}(\mathbb{X}^{\alpha}[\boldsymbol{n}]\mathbf{P}_{\!n}) \\ &+ \mathbf{P}_{\!n}[(\boldsymbol{\Xi}^{\alpha} - \mathrm{div}\,\mathbb{X}^{\alpha})[\boldsymbol{n}] - \mathrm{div}_{\mathcal{S}}(\mathbb{X}^{\alpha}[\boldsymbol{n}]\mathbf{P}_{\!n})] \\ &+ \mathbb{X}^{\alpha}[\boldsymbol{n}][\boldsymbol{n}] \cdot \mathbf{K}^{\top})\,\mathrm{d}a = 0,\end{aligned} \tag{216}$$

and

$$\begin{aligned}\int\limits_{\mathcal{S}} (\boldsymbol{m}_{\mathcal{S}} + \boldsymbol{m}_{\mathcal{S}^*} &+ 2\,\mathrm{ax}\,(\mathrm{skw}\,(\mathrm{div}_{\mathcal{S}}(\mathbf{T}[\boldsymbol{n}]\mathbf{P}_{\!n}))^{\top}) \\ &+ 2\,\mathrm{ax}\,(\mathrm{skw}\,((\mathbb{T} - \mathrm{div}\,\mathbf{T})[\boldsymbol{n}]\mathbf{P}_{\!n})^{\top}) - 2\,\mathrm{ax}\,(\mathrm{skw}\,(\mathrm{div}_{\mathcal{S}}(\mathbf{T}[\boldsymbol{n}]\mathbf{P}_{\!n})\mathbf{P}_{\!n})^{\top}) \\ &+ 2\,\mathrm{ax}\,(\mathrm{skw}\,(\mathbf{T}[\boldsymbol{n}][\boldsymbol{n}]\mathbf{K}^{\top})^{\top}))\,\mathrm{d}a = \mathbf{0},\end{aligned} \tag{217}$$

and by localizing these expressions, we arrive at the following jump conditions across an open surface $\mathcal{S}$,

$$\begin{aligned}\varpi^{\alpha}_{\mathcal{S}} + \varpi^{\alpha}_{\mathcal{S}^*} = &- \mathbf{P}_{\!n}[(\boldsymbol{\Xi}^{\alpha} - \mathrm{div}\,\mathbb{X}^{\alpha})[\boldsymbol{n}]] \\ &- (\boldsymbol{n} \otimes \boldsymbol{n})[\mathrm{div}_{\mathcal{S}}(\mathbb{X}^{\alpha}[\boldsymbol{n}]\mathbf{P}_{\!n})] - \mathbb{X}^{\alpha}[\boldsymbol{n}][\boldsymbol{n}] \cdot \mathbf{K}^{\top},\end{aligned} \tag{218}$$

and

$$\begin{aligned}\boldsymbol{m}_{\mathcal{S}} + \boldsymbol{m}_{\mathcal{S}^*} = &-2\,\mathrm{ax}\,(\mathrm{skw}\,((\mathbb{T} - \mathrm{div}\,\mathbf{T})[\boldsymbol{n}]\mathbf{P}_{\!n})^{\top}) \\ &- 2\,\mathrm{ax}\,(\mathrm{skw}\,(\mathrm{div}_{\mathcal{S}}(\mathbf{T}[\boldsymbol{n}]\mathbf{P}_{\!n})(\boldsymbol{n} \otimes \boldsymbol{n}))^{\top}) - 2\,\mathrm{ax}\,(\mathrm{skw}\,(\mathbf{T}[\boldsymbol{n}][\boldsymbol{n}]\mathbf{K}^{\top})^{\top}).\end{aligned} \tag{219}$$

Note that $-2\,\mathrm{ax}\,(\mathrm{skw}\,(\mathrm{div}_{\mathcal{S}}(\mathbf{T}[\boldsymbol{n}]\mathbf{P}_{\!n})(\boldsymbol{n} \otimes \boldsymbol{n}))^{\top}) = \boldsymbol{n} \times \mathrm{div}_{\mathcal{S}}(\mathbf{T}[\boldsymbol{n}]\mathbf{P}_{\!n})[\boldsymbol{n}]$.

In view of the expressions for the αth surface-couple microtraction $(191)_1$ and the surface-couple traction $(192)_1$, we also have that

$$\varpi^{\alpha}_{\mathcal{S}^*}(\boldsymbol{x}, t; \boldsymbol{n}) = -(\boldsymbol{\Xi}^{\alpha} - \mathrm{div}\,\mathbb{X}^{\alpha})[\boldsymbol{n}], \tag{220}$$

and

$$\boldsymbol{m}_{\mathcal{S}^*}(\boldsymbol{x}, t; \boldsymbol{n}) = -2\,\mathrm{ax}\,(\mathrm{skw}\,((\mathbb{T} - \mathrm{div}\,\mathbb{T})[\boldsymbol{n}])^{\top}). \tag{221}$$

6 The Principle of Virtual Powers on $\overline{\mathcal{S}} \subset \overline{\mathcal{P}}$

The principle of virtual powers may be postulated for an arbitrary surface $\overline{\mathcal{S}}$ while neglecting surface and curve inertial components for an open surface $\mathcal{S} \subset \mathcal{P} \subseteq \mathcal{B}$ as follows.

Postulate 5 (Principle of virtual powers for a nonsmooth open surface) *Neglecting surface and curve inertial components, the principle of virtual powers on a nonsmooth open surface $\overline{\mathcal{S}}$ states that the following balances of virtual powers*

$$\mathcal{V}^{\flat}_{\mathrm{int}}(\overline{\mathcal{S}}; \{\chi^{\alpha}\}^{n}_{\alpha=1}) = \mathcal{V}^{\flat}_{\mathrm{ext}}(\overline{\mathcal{S}}; \{\chi^{\alpha}\}^{n}_{\alpha=1}), \tag{222}$$

and

$$\mathcal{V}^{\sharp}_{\mathrm{int}}(\overline{\mathcal{S}}; \boldsymbol{\chi}) = \mathcal{V}^{\sharp}_{\mathrm{ext}}(\overline{\mathcal{S}}; \boldsymbol{\chi}), \tag{223}$$

hold for any arbitrary control surface $\overline{\mathcal{S}} \subset \mathcal{B}$, time t, and any choice of smooth admissible scalars χ^{α} and vector $\boldsymbol{\chi}$ virtual fields. For the scalar virtual fields, the internal virtual power $\mathcal{V}^{\flat}_{\mathrm{int}}$ is

$$\begin{aligned}\mathcal{V}^{\flat}_{\mathrm{int}}(\overline{\mathcal{S}}; \{\chi^{\alpha}\}^{n}_{\alpha=1}) := \sum_{\alpha=1}^{n} \Bigg\{ & \int\limits_{\overline{\mathcal{S}}} \boldsymbol{\lambda}^{\alpha}_{\mathcal{S}} \cdot \mathrm{grad}_{\mathcal{S}}\chi^{\alpha}\,\mathrm{d}a \\ & + \int\limits_{\overline{\mathcal{C}}} \boldsymbol{\lambda}^{\alpha}_{\mathcal{C}} \cdot (\partial_{\sigma}\chi^{\alpha}\boldsymbol{\sigma})\,\mathrm{d}\sigma + \int\limits_{\overline{\partial\mathcal{S}}} \boldsymbol{\lambda}^{\alpha}_{\partial\mathcal{S}} \cdot (\partial_{\sigma}\chi^{\alpha}\boldsymbol{\sigma})\,\mathrm{d}\sigma \\ & - \int\limits_{\overline{\mathcal{S}}} (\xi^{\alpha}_{\mathcal{S}^*}\chi^{\alpha} + {}^{2}\xi^{\alpha}_{\mathcal{S}^*}\partial_{n}\chi^{\alpha} + {}^{3}\xi^{\alpha}_{\mathcal{S}^*}\partial^{2}_{n}\chi^{\alpha})\,\mathrm{d}a \Bigg\},\end{aligned} \tag{224}$$

and the external virtual power $\mathcal{V}^{\flat}_{\mathrm{ext}}$ is

$$
\begin{aligned}
\mathcal{V}^{\flat}_{\text{ext}}(\overline{\mathcal{S}};\{\chi^\alpha\}_{\alpha=1}^n) := \sum_{\alpha=1}^{n}\Bigg\{ &\int_{\overline{\mathcal{S}}} (\xi^\alpha_{\mathcal{S}}\chi^\alpha + {}^2\xi^\alpha_{\mathcal{S}}\partial_n\chi^\alpha + {}^3\xi^\alpha_{\mathcal{S}}\partial^2_n\chi^\alpha)\,\mathrm{d}a \\
&+ \int_{\overline{\partial\mathcal{S}}} (\xi^\alpha_{\partial\mathcal{S}}\chi^\alpha + {}^2_n\xi^\alpha_{\partial\mathcal{S}}\partial_n\chi^\alpha + {}^2_\nu\xi^\alpha_{\partial\mathcal{S}}\partial_\nu\chi^\alpha)\,\mathrm{d}\sigma \\
&+ \int_{\overline{\mathcal{C}}} (\xi^\alpha_{\mathcal{C}}\chi^\alpha + {}^2_n\xi^\alpha_{\mathcal{C}}\partial_n\chi^\alpha + {}^2_\nu\xi^\alpha_{\mathcal{C}}\partial_\nu\chi^\alpha)\,\mathrm{d}\sigma \\
&+ \xi^\alpha_{\partial^2\mathcal{S}}\chi^\alpha|_{\partial^2\mathcal{S}} + \xi^\alpha_{\partial\mathcal{C}}\chi^\alpha|_{\partial\mathcal{C}} + \xi^\alpha_{\mathcal{O}}\chi^\alpha|_{\mathcal{O}}\Bigg\}.
\end{aligned}
\tag{225}
$$

whereas for the vector virtual field, the internal virtual power $\mathcal{V}^\sharp_{\text{int}}$ *is*

$$
\begin{aligned}
\mathcal{V}^\sharp_{\text{int}}(\overline{\mathcal{S}};\boldsymbol{\chi}) := &\int_{\overline{\mathcal{S}}} \mathbf{Z}_{\mathcal{S}} : \text{grad}_{\mathcal{S}}\boldsymbol{\chi}\,\mathrm{d}a, \\
&+ \int_{\overline{\mathcal{C}}} \mathbf{Z}_{\mathcal{C}} : (\partial_\sigma\boldsymbol{\chi}\otimes\boldsymbol{\sigma})\,\mathrm{d}\sigma + \int_{\overline{\partial\mathcal{S}}} \mathbf{Z}_{\partial\mathcal{S}} : (\partial_\sigma\boldsymbol{\chi}\otimes\boldsymbol{\sigma})\,\mathrm{d}\sigma \\
&- \int_{\overline{\mathcal{S}}} (\boldsymbol{t}_{\mathcal{S}*}\cdot\boldsymbol{\chi} + {}^2\boldsymbol{t}_{\mathcal{S}*}\cdot\partial_n\boldsymbol{\chi} + {}^3\boldsymbol{t}_{\mathcal{S}*}\cdot\partial^2_n\boldsymbol{\chi})\,\mathrm{d}a
\end{aligned}
\tag{226}
$$

and the external virtual power $\mathcal{V}^\sharp_{\text{ext}}$ *is*

$$
\begin{aligned}
\mathcal{V}^\sharp_{\text{ext}}(\overline{\mathcal{S}};\boldsymbol{\chi}) := &\int_{\overline{\mathcal{S}}} (\boldsymbol{t}_{\mathcal{S}}\cdot\boldsymbol{\chi} + {}^2\boldsymbol{t}_{\mathcal{S}}\cdot\partial_n\boldsymbol{\chi} + {}^3\boldsymbol{t}_{\mathcal{S}}\cdot\partial^2_n\boldsymbol{\chi})\,\mathrm{d}a \\
&+ \int_{\overline{\partial\mathcal{S}}} (\boldsymbol{t}_{\partial\mathcal{S}}\cdot\boldsymbol{\chi} + {}^2_n\boldsymbol{t}_{\partial\mathcal{S}}\cdot\partial_n\boldsymbol{\chi} + {}^2_\nu\boldsymbol{t}_{\partial\mathcal{S}}\cdot\partial_\nu\boldsymbol{\chi})\,\mathrm{d}\sigma \\
&+ \int_{\overline{\mathcal{C}}} (\boldsymbol{t}_{\mathcal{C}}\cdot\boldsymbol{\chi} + {}^2_n\boldsymbol{t}_{\mathcal{C}}\cdot\partial_n\boldsymbol{\chi} + {}^2_\nu\boldsymbol{t}_{\mathcal{C}}\cdot\partial_\nu\boldsymbol{\chi})\,\mathrm{d}\sigma \\
&+ \boldsymbol{t}_{\partial^2\mathcal{S}}\cdot\boldsymbol{\chi}|_{\partial^2\mathcal{S}} + \boldsymbol{t}_{\partial\mathcal{C}}\cdot\boldsymbol{\chi}|_{\partial\mathcal{C}} + \boldsymbol{t}_{\mathcal{O}}\cdot\boldsymbol{\chi}|_{\mathcal{O}}.
\end{aligned}
\tag{227}
$$

Here, $\boldsymbol{\lambda}^\alpha_{\mathcal{S}}$ and $\mathbf{Z}_{\mathcal{S}}$ are the αth surface microstress and the surface stress, respectively, $\boldsymbol{\lambda}^\alpha_{\mathcal{C}}$ and $\boldsymbol{\lambda}^\alpha_{\partial\mathcal{S}}$ are, respectively, the αth junction-edge and αth boundary-edge microstresses, whereas $\mathbf{Z}_{\mathcal{C}}$ and $\mathbf{Z}_{\partial\mathcal{S}}$ are, respectively, the junction-edge and boundary-edge stresses. Also, ${}^2\xi^\alpha_{\mathcal{S}*}$, ${}^3\xi^\alpha_{\mathcal{S}*}$, ${}^2\boldsymbol{t}_{\mathcal{S}*}$, and ${}^3\boldsymbol{t}_{\mathcal{S}*}$ are, respectively, the αth surface hypermicrotraction power conjugate to $\partial_n\chi$, the αth surface supra-hypermicrotraction power conjugate to $\partial_n^2\chi$, the surface hypertraction power conjugate to $\partial_n\boldsymbol{\chi}$, and the surface supra-hypertraction power conjugate to $\partial_n^2\boldsymbol{\chi}$. All these quantities are developed on the opposite side of $\mathcal{S}$, that is, $\mathcal{S}^*$. Moreover, the αth surface microstress $\boldsymbol{\lambda}_{\mathcal{S}}^{\alpha}$, the αth junction-edge microstress $\boldsymbol{\lambda}_{\mathcal{C}}^{\alpha}$, the αth boundary-edge microstress $\boldsymbol{\lambda}_{\partial\mathcal{S}}^{\alpha}$ the surface stress $\mathbf{Z}_{\mathcal{S}}$, the junction-edge stress $\mathbf{Z}_{\mathcal{C}}$, the boundary-edge stress $\mathbf{Z}_{\partial\mathcal{S}}$ along with the hyperfields ${}^2\xi_{\mathcal{S}^*}^{\alpha}$, ${}^3\xi_{\mathcal{S}^*}^{\alpha}$, ${}^2\boldsymbol{t}_{\mathcal{S}^*}$, and ${}^3\boldsymbol{t}_{\mathcal{S}^*}$ are the only unknown fields arising in (222) and (223).

Next, consider the jump $\xi_{\mathcal{S}}^{\alpha} + \xi_{\mathcal{S}^*}^{\alpha}$ given in expression (211). The expressions for the remaining known fields are given in (113)–(119). Replacing these expressions in the external virtual power (225), with the gradient theorem for nonsmooth closed $(82)_1$ and open $(83)_1$ curves and the surface divergence theorem for nonsmooth open surfaces $(77)_1$, we arrive at

$$\begin{aligned}
&\sum_{\alpha=1}^{n}\Bigg\{\int_{\overline{\mathcal{S}}}(\boldsymbol{\lambda}_{\mathcal{S}}^{\alpha} - \mathbf{P}_{\!n}[(\boldsymbol{\Xi}^{\alpha} + \operatorname{div}\mathbb{X}^{\alpha})[\boldsymbol{n}]] \\
&\qquad\qquad + \mathbf{P}_{\!n}[\operatorname{div}_{\mathcal{S}}(\mathbb{X}^{\alpha}[\boldsymbol{n}]\mathbf{P}_{\!n})] - \mathbb{X}^{\alpha}[\boldsymbol{n}][\boldsymbol{n}]\cdot\mathbf{K}^{\top})\cdot\operatorname{grad}_{\mathcal{S}}\chi^{\alpha}\,\mathrm{d}a \\
&\quad + \int_{\overline{\mathcal{C}}}(\boldsymbol{\lambda}_{\mathcal{C}}^{\alpha} - \{\!\!\{\mathbb{X}^{\alpha}[\boldsymbol{n}][\boldsymbol{\nu}]\}\!\!\})\cdot\partial_\sigma\chi^{\alpha}\boldsymbol{\sigma}\,\mathrm{d}\sigma + \int_{\overline{\partial\mathcal{S}}}(\boldsymbol{\lambda}_{\partial\mathcal{S}}^{\alpha} - \mathbb{X}^{\alpha}[\boldsymbol{n}][\boldsymbol{\nu}])\cdot\partial_\sigma\chi^{\alpha}\boldsymbol{\sigma}\,\mathrm{d}\sigma\Bigg\} \\
&\qquad = \sum_{\alpha=1}^{n}\Bigg\{\int_{\overline{\mathcal{S}}}\big(({}^2\xi_{\mathcal{S}}^{\alpha} + {}^2\xi_{\mathcal{S}^*}^{\alpha})\partial_n\chi^{\alpha} + ({}^3\xi_{\mathcal{S}}^{\alpha} + {}^3\xi_{\mathcal{S}^*}^{\alpha})\partial_n^2\chi^{\alpha}\big)\,\mathrm{d}a\Bigg\}.
\end{aligned} \tag{228}$$

By localization, we obtain the αth surface microstress

$$\boldsymbol{\lambda}_{\mathcal{S}}^{\alpha} = \mathbf{P}_{\!n}[(\boldsymbol{\Xi}^{\alpha} + \operatorname{div}\mathbb{X}^{\alpha})[\boldsymbol{n}]] - \mathbf{P}_{\!n}[\operatorname{div}_{\mathcal{S}}(\mathbb{X}^{\alpha}[\boldsymbol{n}]\mathbf{P}_{\!n})] + \mathbb{X}^{\alpha}[\boldsymbol{n}][\boldsymbol{n}]\cdot\mathbf{K}^{\top}, \tag{229}$$

the αth junction-edge microstress

$$\boldsymbol{\lambda}_{\mathcal{C}}^{\alpha} = \{\!\!\{\mathbb{X}^{\alpha}[\boldsymbol{n}][\boldsymbol{\nu}]\}\!\!\}, \tag{230}$$

the αth boundary-edge microstress

$$\boldsymbol{\lambda}_{\partial\mathcal{S}}^{\alpha} = \mathbb{X}^{\alpha}[\boldsymbol{n}][\boldsymbol{\nu}], \tag{231}$$

and the jump conditions for the αth surface hypermicrotraction and the αth surface supra-hypermicrotraction

$$\begin{cases} {}^2\xi_{\mathcal{S}}^{\alpha} + {}^2\xi_{\mathcal{S}^*}^{\alpha} = 0, \\ {}^3\xi_{\mathcal{S}}^{\alpha} + {}^3\xi_{\mathcal{S}^*}^{\alpha} = 0. \end{cases} \tag{232}$$

Moreover, with the αth surface hypermicrotraction (114) and the αth surface supra-hypermicrotraction (115), we have that

$$\begin{cases} {}^2\xi^{\alpha}_{\mathcal{S}^*} = -\boldsymbol{n} \cdot (\Xi^{\alpha} + \operatorname{div}\mathbb{X}^{\alpha})[\boldsymbol{n}] + 2\operatorname{div}_{\mathcal{S}}(\mathbf{P}_{\!n}[\mathbb{X}^{\alpha}[\boldsymbol{n}][\boldsymbol{n}]]) + \mathbb{X}^{\alpha}[\boldsymbol{n}] : \mathbf{K}, \\ {}^3\xi^{\alpha}_{\mathcal{S}^*} = -\boldsymbol{n} \cdot \mathbb{X}^{\alpha}[\boldsymbol{n}][\boldsymbol{n}]. \end{cases} \tag{233}$$

Analogously, consider the jump $\boldsymbol{t}_{\mathcal{S}} + \boldsymbol{t}_{\mathcal{S}^*}$ given in expression (212). The expressions for the remaining known fields are given in (129)–(135). Replacing these expressions in the external virtual power (227), with the gradient theorem for nonsmooth closed $(82)_1$ and open $(83)_1$ curves and the surface divergence theorem for nonsmooth open surfaces $(77)_1$, we arrive at

$$\int_{\overline{\mathcal{S}}} (\mathbf{Z}_{\mathcal{S}} - (\mathbb{T} - \operatorname{div}\mathbf{T})[\boldsymbol{n}]\mathbf{P}_{\!n} + \operatorname{div}_{\mathcal{S}}(\mathbf{T}[\boldsymbol{n}]\mathbf{P}_{\!n})\mathbf{P}_{\!n} - \mathbf{T}[\boldsymbol{n}][\boldsymbol{n}]\mathbf{K}^{\top}) : \operatorname{grad}_{\mathcal{S}}\chi \, da =$$

$$\int_{\overline{\mathcal{S}}} \big(({}^2\boldsymbol{t}_{\mathcal{S}} + {}^2\boldsymbol{t}_{\mathcal{S}^*}) \cdot \partial_n \chi + ({}^3\boldsymbol{t}_{\mathcal{S}} + {}^3\boldsymbol{t}_{\mathcal{S}^*}) \cdot \partial_n^2 \chi\big) \, da. \tag{234}$$

By localization, we obtain the surface stress

$$\mathbf{Z}_{\mathcal{S}} = (\mathbb{T} - \operatorname{div}\mathbf{T})[\boldsymbol{n}]\mathbf{P}_{\!n} - \operatorname{div}_{\mathcal{S}}(\mathbf{T}[\boldsymbol{n}]\mathbf{P}_{\!n})\mathbf{P}_{\!n} + \mathbf{T}[\boldsymbol{n}][\boldsymbol{n}]\mathbf{K}^{\top}, \tag{235}$$

the junction-edge stress

$$\mathbf{Z}_{\mathcal{C}} = \{\!\{\mathbf{T}[\boldsymbol{n}][\boldsymbol{\nu}]\}\!\}, \tag{236}$$

the boundary-edge stress

$$\mathbf{Z}_{\partial\mathcal{S}} = \mathbf{T}[\boldsymbol{n}][\boldsymbol{\nu}], \tag{237}$$

and the jump conditions for the surface hypertraction and surface supra-hypertraction

$$\begin{cases} {}^2\boldsymbol{t}_{\mathcal{S}} + {}^2\boldsymbol{t}_{\mathcal{S}^*} = \mathbf{0}, \\ {}^3\boldsymbol{t}_{\mathcal{S}} + {}^3\boldsymbol{t}_{\mathcal{S}^*} = \mathbf{0}. \end{cases} \tag{238}$$

Moreover, considering the surface hypermicrotraction (130) and the surface supra-hypermicrotraction (131), we have that

$$\begin{cases} {}^2\boldsymbol{t}_{\mathcal{S}^*} = -(\mathbb{T} - \operatorname{div}\mathbf{T})[\boldsymbol{n}][\boldsymbol{n}] + 2\operatorname{div}_{\mathcal{S}}(\mathbf{T}[\boldsymbol{n}][\boldsymbol{n}]\mathbf{P}_{\!n}) + \mathbf{T}[\boldsymbol{n}][\mathbf{K}], \\ {}^3\boldsymbol{t}_{\mathcal{S}^*} = -\mathbf{T}[\boldsymbol{n}][\boldsymbol{n}][\boldsymbol{n}]. \end{cases} \tag{239}$$

Remark 3 *(On the frame indifference for the principle of virtual powers on surfaces)* As we have seen, the frame indifference principle given in Postulate 2 requires that the internal power must be invariant under changes of frame. The first consequence, given in expression (157), is that the stress $\mathbf{T}$ should be symmetric. Another immediate (second) consequence is that the external virtual power is invariant as well.

Likewise, one may stipulate the need and wonder about the consequences of the frame indifference for the principle of virtual power on surfaces. Although we may reckon, without much formalism, that the virtual power balance on surfaces is invariant, we find that there is no need for such a requirement.

Furthermore, from the second consequence above, given that the external virtual power is frame indifferent on a control volume, the external power on a surface $\overline{\mathcal{S}}$ is invariant as well, and so is the internal virtual power on $\overline{\mathcal{S}}$.

Synopsis, chapter 4

Microkinetics

(i) Boundary-edge microtraction on $\partial\mathcal{S}$:

$$\xi^{\alpha}_{\partial\mathcal{S}} = \boldsymbol{\nu}\cdot((\boldsymbol{\Xi}^{\alpha} - \operatorname{div}\mathbb{X}^{\alpha})[\boldsymbol{n}] - \operatorname{div}_{\mathcal{S}}(\mathbb{X}^{\alpha}[\boldsymbol{n}]\mathbf{P}_{\!n}) + \mathbb{X}^{\alpha}[\boldsymbol{n}][\boldsymbol{n}]\cdot\mathbf{K}^{\top}) - \partial_{\sigma}(\boldsymbol{\sigma}\cdot\mathbb{X}^{\alpha}[\boldsymbol{n}][\boldsymbol{\nu}])$$

(ii) Surface-couple microtraction on $\mathcal{S}$:

$$\boldsymbol{\varpi}^{\alpha}_{\mathcal{S}} = (\boldsymbol{n}\otimes\boldsymbol{n})[(\boldsymbol{\Xi}^{\alpha} - \operatorname{div}\mathbb{X}^{\alpha})[\boldsymbol{n}] - \operatorname{div}_{\mathcal{S}}(\mathbb{X}^{\alpha}[\boldsymbol{n}]\mathbf{P}_{\!n})] - \mathbb{X}^{\alpha}[\boldsymbol{n}][\boldsymbol{n}]\cdot\mathbf{K}^{\top}$$

(iii) Boundary-edge-couple microtraction on $\mathcal{C}$:

$$\boldsymbol{\varpi}^{\alpha}_{\partial\mathcal{S}} = \mathbf{P}_{\!\sigma}[\mathbb{X}^{\alpha}[\boldsymbol{n}][\boldsymbol{\nu}]]$$

(iv) Junction-edge-couple microtraction on $\mathcal{C}$:

$$\boldsymbol{\varpi}^{\alpha}_{\mathcal{C}} = \{\!\!\{\mathbf{P}_{\!\sigma}[\mathbb{X}^{\alpha}[\boldsymbol{n}][\boldsymbol{\nu}]]\}\!\!\}$$

(v) Surface microtraction jump condition across $\mathcal{S}$:

$$\xi^{\alpha}_{\mathcal{S}} + \xi^{\alpha}_{\mathcal{S}^*} = -\operatorname{div}_{\mathcal{S}}(\mathbf{P}_{\!n}[(\boldsymbol{\Xi}^{\alpha} + \operatorname{div}\mathbb{X}^{\alpha})[\boldsymbol{n}]]) + \operatorname{div}_{\mathcal{S}}(\mathbf{P}_{\!n}[\operatorname{div}_{\mathcal{S}}(\mathbb{X}^{\alpha}[\boldsymbol{n}]\mathbf{P}_{\!n})]) - \operatorname{div}_{\mathcal{S}}(\mathbb{X}^{\alpha}[\boldsymbol{n}][\boldsymbol{n}]\cdot\mathbf{K}^{\top})$$

(vi) Surface-couple microtraction jump condition across $\mathcal{S}$:

$$\boldsymbol{\varpi}^{\alpha}_{\mathcal{S}} + \boldsymbol{\varpi}^{\alpha}_{\mathcal{S}^*} = -\,\mathbf{P}_{\!n}[(\boldsymbol{\Xi}^{\alpha} - \operatorname{div}\mathbb{X}^{\alpha})[\boldsymbol{n}]] - (\boldsymbol{n}\otimes\boldsymbol{n})[\operatorname{div}_{\mathcal{S}}(\mathbb{X}^{\alpha}[\boldsymbol{n}]\mathbf{P}_{\!n})] - \mathbb{X}^{\alpha}[\boldsymbol{n}][\boldsymbol{n}]\cdot\mathbf{K}^{\top}$$

(vii) Surface microtraction on $\mathcal{S}^*$:

$$\xi^{\alpha}_{\mathcal{S}^*} = -(\boldsymbol{\xi}^{\alpha} - \operatorname{div}\boldsymbol{\Xi}^{\alpha} + \operatorname{div}^2\mathbb{X}^{\alpha})\cdot\boldsymbol{n}$$

(viii) Surface-couple microtraction on $\mathcal{S}^*$:

$$\boldsymbol{\varpi}^{\alpha}_{\mathcal{S}^*} = -(\boldsymbol{\Xi}^{\alpha} - \operatorname{div}\mathbb{X}^{\alpha})[\boldsymbol{n}]$$

(ix) Surface balance of virtual powers on $\overline{\mathcal{S}}$ for scalar virtual fields:

$$
\underbrace{\begin{aligned}
\sum_{\alpha=1}^{n} \Bigg\{ \int_{\overline{\mathcal{S}}} \boldsymbol{\lambda}_{\mathcal{S}}^{\alpha} \cdot \operatorname{grad}_{\mathcal{S}} \chi^{\alpha} \, \mathrm{d}a
&+ \int_{\overline{\mathcal{C}}} \boldsymbol{\lambda}_{\mathcal{C}}^{\alpha} \cdot (\partial_{\sigma} \chi^{\alpha} \boldsymbol{\sigma}) \, \mathrm{d}\sigma + \int_{\overline{\partial \mathcal{S}}} \boldsymbol{\lambda}_{\partial \mathcal{S}}^{\alpha} \cdot (\partial_{\sigma} \chi^{\alpha} \boldsymbol{\sigma}) \, \mathrm{d}\sigma \\
&- \int_{\overline{\mathcal{S}}} (\xi_{\mathcal{S}^*}^{\alpha} \chi^{\alpha} + {}^{2}\xi_{\mathcal{S}^*}^{\alpha} \partial_n \chi^{\alpha} + {}^{3}\xi_{\mathcal{S}^*}^{\alpha} \partial_n^2 \chi^{\alpha}) \, \mathrm{d}a \Bigg\}
\end{aligned}}_{\mathcal{V}_{\mathrm{int}}^{\flat}(\overline{\mathcal{S}}; \{\chi^{\alpha}\}_{\alpha=1}^{n})} =
$$

$$
\underbrace{\begin{aligned}
\sum_{\alpha=1}^{n} \Bigg\{ \int_{\overline{\mathcal{S}}} (\xi_{\mathcal{S}}^{\alpha} \chi^{\alpha} + {}^{2}\xi_{\mathcal{S}}^{\alpha} \partial_n \chi^{\alpha} + {}^{3}\xi_{\mathcal{S}}^{\alpha} \partial_n^2 \chi^{\alpha}) \, \mathrm{d}a
&+ \int_{\overline{\partial \mathcal{S}}} (\xi_{\partial \mathcal{S}}^{\alpha} \chi^{\alpha} + {}_{n}^{2}\xi_{\partial \mathcal{S}}^{\alpha} \partial_n \chi^{\alpha} + {}_{\nu}^{2}\xi_{\partial \mathcal{S}}^{\alpha} \partial_{\nu} \chi^{\alpha}) \, \mathrm{d}\sigma \\
&+ \int_{\overline{\mathcal{C}}} (\xi_{\mathcal{C}}^{\alpha} \chi^{\alpha} + {}_{n}^{2}\xi_{\mathcal{C}}^{\alpha} \partial_n \chi^{\alpha} + {}_{\nu}^{2}\xi_{\mathcal{C}}^{\alpha} \partial_{\nu} \chi^{\alpha}) \, \mathrm{d}\sigma \\
&+ \xi_{\partial^2 \mathcal{S}}^{\alpha} \chi^{\alpha}|_{\partial^2 \mathcal{S}} + \xi_{\partial \mathcal{C}}^{\alpha} \chi^{\alpha}|_{\partial \mathcal{C}} + \xi_{\mathcal{O}}^{\alpha} \chi^{\alpha}|_{\mathcal{O}} \Bigg\}
\end{aligned}}_{\mathcal{V}_{\mathrm{ext}}^{\flat}(\overline{\mathcal{S}}; \{\chi^{\alpha}\}_{\alpha=1}^{n})}
$$

(x) Surface microstress on $\mathcal{S}$:

$$
\boldsymbol{\lambda}_{\mathcal{S}}^{\alpha} = \mathbf{P}_{n}[(\boldsymbol{\Xi}^{\alpha} + \operatorname{div} \mathbb{X}^{\alpha})[\boldsymbol{n}]] - \mathbf{P}_{n}[\operatorname{div}_{\mathcal{S}}(\mathbb{X}^{\alpha}[\boldsymbol{n}]\mathbf{P}_{n})] + \mathbb{X}^{\alpha}[\boldsymbol{n}][\boldsymbol{n}] \cdot \mathbf{K}^{\top}
$$

(xi) Junction-edge microstress on $\mathcal{C}$:

$$
\boldsymbol{\lambda}_{\mathcal{C}}^{\alpha} = \{\!\!\{ \mathbb{X}^{\alpha}[\boldsymbol{n}][\boldsymbol{\nu}] \}\!\!\}
$$

(xii) Boundary-edge microstress on $\partial \mathcal{S}$:

$$
\boldsymbol{\lambda}_{\partial \mathcal{S}}^{\alpha} = \mathbb{X}^{\alpha}[\boldsymbol{n}][\boldsymbol{\nu}]
$$

(xiii) Surface hypermicrotraction and surface supra-hypermicrotraction jump conditions across $\mathcal{S}$:

$$
\begin{cases}
{}^{2}\xi_{\mathcal{S}}^{\alpha} + {}^{2}\xi_{\mathcal{S}^*}^{\alpha} = 0 \\
{}^{3}\xi_{\mathcal{S}}^{\alpha} + {}^{3}\xi_{\mathcal{S}^*}^{\alpha} = 0
\end{cases}
$$

(xiv) Surface hypermicrotraction and surface supra-hypermicrotraction on $\mathcal{S}^*$:

$$\begin{cases} {}^2\xi^\alpha_{\mathcal{S}^*} = -\boldsymbol{n}\cdot(\Xi^\alpha + \operatorname{div}\mathbb{X}^\alpha)[\boldsymbol{n}] + 2\operatorname{div}_{\mathcal{S}}(\mathbf{P}_{\!n}[\mathbb{X}^\alpha[\boldsymbol{n}][\boldsymbol{n}]]) + \mathbb{X}^\alpha[\boldsymbol{n}] : \mathbf{K} \\ {}^3\xi^\alpha_{\mathcal{S}^*} = -\boldsymbol{n}\cdot\mathbb{X}^\alpha[\boldsymbol{n}][\boldsymbol{n}] \end{cases}$$

Kinetics

(i) Boundary-edge traction on $\partial\mathcal{S}$*:*

$$\begin{aligned} \boldsymbol{t}_{\partial\mathcal{S}} = {} & (\mathbb{T} - \operatorname{div}\mathbf{T})[\boldsymbol{n}][\boldsymbol{\nu}] - \operatorname{div}_{\mathcal{S}}((\mathbf{T}[\boldsymbol{n}])\mathbf{P}_{\!n})[\boldsymbol{\nu}] \\ & + \mathbf{T}[\boldsymbol{n}][\boldsymbol{n}][\mathbf{K}^\top[\boldsymbol{\nu}]] - \partial_\sigma(\mathbf{T}[\boldsymbol{n}][\boldsymbol{\nu}][\boldsymbol{\sigma}]) \end{aligned}$$

(ii) Surface-couple traction on $\mathcal{S}$*:*

$$\begin{aligned} \boldsymbol{m}_{\mathcal{S}} = {} & \boldsymbol{n}\times((\mathbb{T} - \operatorname{div}\mathbf{T})[\boldsymbol{n}][\boldsymbol{n}] - \operatorname{div}_{\mathcal{S}}(\mathbf{T}[\boldsymbol{n}]\mathbf{P}_{\!n})[\boldsymbol{n}]) \\ & - 2\operatorname{ax}(\operatorname{skw}(\mathbf{T}[\boldsymbol{n}][\boldsymbol{n}]\mathbf{K}^\top)^\top) \end{aligned}$$

(iii) Boundary-edge-couple traction on $\partial\mathcal{S}$*:*

$$\boldsymbol{m}_{\partial\mathcal{S}} = 2\operatorname{ax}(\operatorname{skw}(\mathbf{T}[\boldsymbol{n}][\boldsymbol{\nu}]^\top)) - \boldsymbol{\sigma}\times\mathbf{T}[\boldsymbol{n}][\boldsymbol{\nu}][\boldsymbol{\sigma}]$$

(iv) Junction-edge-couple traction on $\mathcal{C}$*:*

$$\boldsymbol{m}_{\mathcal{C}} = 2\operatorname{ax}(\operatorname{skw}(\{\!\!\{\mathbf{T}[\boldsymbol{n}][\boldsymbol{\nu}]\}\!\!\})^\top) - \boldsymbol{\sigma}\times\{\!\!\{\mathbf{T}[\boldsymbol{n}][\boldsymbol{\nu}][\boldsymbol{\sigma}]\}\!\!\}$$

(v) Surface traction jump condition across $\mathcal{S}$*:*

$$\begin{aligned} \boldsymbol{t}_{\mathcal{S}} + \boldsymbol{t}_{\mathcal{S}^*} = {} & -\operatorname{div}_{\mathcal{S}}((\mathbb{T} - \operatorname{div}\mathbf{T})[\boldsymbol{n}]\mathbf{P}_{\!n})) \\ & + \operatorname{div}_{\mathcal{S}}(\operatorname{div}_{\mathcal{S}}(\mathbf{T}[\boldsymbol{n}]\mathbf{P}_{\!n})\mathbf{P}_{\!n}) - \operatorname{div}_{\mathcal{S}}(\mathbf{T}[\boldsymbol{n}][\boldsymbol{n}]\mathbf{K}^\top) \end{aligned}$$

(vi) Surface-couple traction jump condition across $\mathcal{S}$*:*

$$\begin{aligned} \boldsymbol{m}_{\mathcal{S}} + \boldsymbol{m}_{\mathcal{S}^*} = {} & -2\operatorname{ax}(\operatorname{skw}((\mathbb{T} - \operatorname{div}\mathbf{T})[\boldsymbol{n}]\mathbf{P}_{\!n})^\top) \\ & - 2\operatorname{ax}(\operatorname{skw}(\operatorname{div}_{\mathcal{S}}(\mathbf{T}[\boldsymbol{n}]\mathbf{P}_{\!n})(\boldsymbol{n}\otimes\boldsymbol{n}))^\top) - 2\operatorname{ax}(\operatorname{skw}(\mathbf{T}[\boldsymbol{n}][\boldsymbol{n}]\mathbf{K}^\top)^\top) \end{aligned}$$

(vii) Surface traction on $\mathcal{S}^*$*:*

$$\boldsymbol{t}_{\mathcal{S}^*} = -(\mathbf{T} - \operatorname{div}\mathbb{T} + \operatorname{div}^2\mathbf{T})[\boldsymbol{n}]$$

(viii) Surface-couple traction on $\mathcal{S}^*$*:*

$$\boldsymbol{m}_{\mathcal{S}^*} = -2\operatorname{ax}(\operatorname{skw}((\mathbb{T} - \operatorname{div}\mathbf{T})[\boldsymbol{n}])^\top)$$

(ix) *Surface balance of virtual powers on $\overline{\mathcal{S}}$ for a vector virtual field:*

$$\underbrace{\int_{\overline{\mathcal{S}}} \mathbf{Z}_{\mathcal{S}} : \operatorname{grad}_{\mathcal{S}} \boldsymbol{\chi} \, \mathrm{d}a + \int_{\overline{\mathcal{C}}} \mathbf{Z}_{\mathcal{C}} : (\partial_\sigma \boldsymbol{\chi} \otimes \boldsymbol{\sigma}) \, \mathrm{d}\sigma + \int_{\overline{\partial\mathcal{S}}} \mathbf{Z}_{\partial\mathcal{S}} : (\partial_\sigma \boldsymbol{\chi} \otimes \boldsymbol{\sigma}) \, \mathrm{d}\sigma - \int_{\overline{\mathcal{S}}} \left(\boldsymbol{t}_{\mathcal{S}^*} \cdot \boldsymbol{\chi} + {}^2\boldsymbol{t}_{\mathcal{S}^*} \cdot \partial_n \boldsymbol{\chi} + {}^3\boldsymbol{t}_{\mathcal{S}^*} \cdot \partial_n^2 \boldsymbol{\chi} \right) \mathrm{d}a}_{\mathcal{V}^\sharp_{\mathrm{int}}(\overline{\mathcal{S}};\boldsymbol{\chi})} =$$

$$\underbrace{\begin{aligned} &\int_{\overline{\mathcal{S}}} \left(\boldsymbol{t}_{\mathcal{S}} \cdot \boldsymbol{\chi} + {}^2\boldsymbol{t}_{\mathcal{S}} \cdot \partial_n \boldsymbol{\chi} + {}^3\boldsymbol{t}_{\mathcal{S}} \cdot \partial_n^2 \boldsymbol{\chi} \right) \mathrm{d}a \\ &+ \int_{\overline{\partial\mathcal{S}}} \left(\boldsymbol{t}_{\partial\mathcal{S}} \cdot \boldsymbol{\chi} + {}^2_n\boldsymbol{t}_{\partial\mathcal{S}} \cdot \partial_n \boldsymbol{\chi} + {}^2_\nu\boldsymbol{t}_{\partial\mathcal{S}} \cdot \partial_\nu \boldsymbol{\chi} \right) \mathrm{d}\sigma \\ &+ \int_{\overline{\mathcal{C}}} \left(\boldsymbol{t}_{\mathcal{C}} \cdot \boldsymbol{\chi} + {}^2_n\boldsymbol{t}_{\mathcal{C}} \cdot \partial_n \boldsymbol{\chi} + {}^2_\nu\boldsymbol{t}_{\mathcal{C}} \cdot \partial_\nu \boldsymbol{\chi} \right) \mathrm{d}\sigma \\ &+ \boldsymbol{t}_{\partial^2\mathcal{S}} \cdot \boldsymbol{\chi}|_{\partial^2\mathcal{S}} + \boldsymbol{t}_{\partial\mathcal{C}} \cdot \boldsymbol{\chi}|_{\partial\mathcal{C}} + \boldsymbol{t}_{\mathcal{O}} \cdot \boldsymbol{\chi}|_{\mathcal{O}} \end{aligned}}_{\mathcal{V}^\sharp_{\mathrm{ext}}(\overline{\mathcal{S}};\boldsymbol{\chi})}$$

(x) *Surface stress on $\mathcal{S}$:*

$$\mathbf{Z}_{\mathcal{S}} = (\mathbb{T} - \operatorname{div} \mathbf{T})[\boldsymbol{n}]\mathbf{P}_n - \operatorname{div}_{\mathcal{S}}(\mathbb{T}[\boldsymbol{n}]\mathbf{P}_n)\mathbf{P}_n + \mathbb{T}[\boldsymbol{n}][\boldsymbol{n}]\mathbf{K}^\top$$

(xi) *Junction-edge stress on $\mathcal{C}$:*

$$\mathbf{Z}_{\mathcal{C}} = \{\!\{\mathbb{T}[\boldsymbol{n}][\boldsymbol{\nu}]\}\!\}$$

(xii) *Boundary-edge stress on $\partial\mathcal{S}$:*

$$\mathbf{Z}_{\partial\mathcal{S}} = \mathbb{T}[\boldsymbol{n}][\boldsymbol{\nu}]$$

(xiii) *Surface hypertraction and surface supra-hypertraction jump conditions across $\mathcal{S}$:*

$$\begin{cases} {}^2\boldsymbol{t}_{\mathcal{S}} + {}^2\boldsymbol{t}_{\mathcal{S}^*} = \mathbf{0} \\ {}^3\boldsymbol{t}_{\mathcal{S}} + {}^3\boldsymbol{t}_{\mathcal{S}^*} = \mathbf{0} \end{cases}$$

(xiv) *Surface hypertraction and surface supra-hypertraction on $\mathcal{S}^*$:*

$$\begin{cases} {}^2\boldsymbol{t}_{\mathcal{S}^*} = -(\mathbb{T} - \operatorname{div}\mathbf{T})[\boldsymbol{n}][\boldsymbol{n}] + 2\operatorname{div}_{\mathcal{S}}(\mathbf{T}[\boldsymbol{n}][\boldsymbol{n}]\mathbf{P}_n) + \mathbf{T}[\boldsymbol{n}][\mathbf{K}] \\ {}^3\boldsymbol{t}_{\mathcal{S}^*} = -\mathbf{T}[\boldsymbol{n}][\boldsymbol{n}][\boldsymbol{n}] \end{cases}$$

Part III
Thermodynamics and Constitutive Relations

Chapter 5
Thermodynamics

In this chapter, we restrict attention to the first and second law of thermodynamics expressed on a material part and a control volume. From these postulates, we obtain the pointwise version of the entropy imbalance and free-energy imbalance. These imbalances are useful tools to propose thermodynamically consistent constitutive relations. We then present and extend ideas of Larché–Cahn about derivatives on constrained spaces for the case in which the phase fields are related by an additional algebraic constraint. These ideas are convenient when the phase fields represent conserved species. Lastly, focusing on isotropic tensors, we conclude this chapter with the representation theorem for tensors of order four, six, and eight based on the Clebsch–Gordan decomposition theorem.

1 First and Second Laws of Thermodynamics for a Spatial Part $\mathcal{P}_\tau$

We now postulate the first two laws of thermodynamics for a continuum consisting of balance of energy and an entropy imbalance, which is frequently referred to as the Clausius–Duhem inequality. Let $\mathcal{P}_\tau$ be an arbitrary part that advects with the body. Following Truesdell & Noll [53, §79], these laws have the respective forms[1]

[1] The idea of using a virtual power principle to generate an appropriate form of the external power expenditure in the energy balance was originated by Gurtin [42, §6].

L. Espath, *Mechanics and Geometry of Enriched Continua*,
https://doi.org/10.1007/978-3-031-28934-7_5

$$\begin{cases} \overline{\int\limits_{\mathcal{P}_\tau} \varrho(\varepsilon + \tfrac{1}{2}|\boldsymbol{v}|^2)\,\mathrm{d}v_\tau}^{\boldsymbol{\cdot}} = \mathcal{W}^{\mathrm{conv}}_{\mathrm{ext}}(\mathcal{P}_\tau) + \int\limits_{\mathcal{P}_\tau} r\,\mathrm{d}v_\tau - \int\limits_{\partial\mathcal{P}_\tau} \boldsymbol{q}\cdot\boldsymbol{n}\,\mathrm{d}a_\tau, \\ \overline{\int\limits_{\mathcal{P}_\tau} \varrho\eta\,\mathrm{d}v_\tau}^{\boldsymbol{\cdot}} \geq \int\limits_{\mathcal{P}_\tau} \frac{r}{\vartheta}\,\mathrm{d}v_\tau - \int\limits_{\partial\mathcal{P}_\tau} \frac{\boldsymbol{q}}{\vartheta}\cdot\boldsymbol{n}\,\mathrm{d}a_\tau, \end{cases} \tag{240}$$

where ε and η represent the internal-energy density and entropy density, $\boldsymbol{q}$ is the heat flux, r is the heat supply, and $\vartheta > 0$ is the absolute temperature. Note that $\mathcal{W}^{\mathrm{conv}}_{\mathrm{ext}}$ does not include inertial power and is regarded as the *conventional external power* which differs from the *generalized external power* previously presented. These two laws of thermodynamics (240) with (103) and (105) may be rewritten, in the context of this study, as

$$\begin{cases} \overline{\int\limits_{\mathcal{P}_\tau} \varrho(\varepsilon + \tfrac{1}{2}|\boldsymbol{v}|^2)\,\mathrm{d}v_\tau}^{\boldsymbol{\cdot}} = \int\limits_{\mathcal{P}_\tau} \Big(\boldsymbol{b}^{\mathrm{ni}}\cdot\boldsymbol{v} + \sum_{\alpha=1}^{n} \gamma^\alpha\dot{\varphi}^\alpha + r\Big)\,\mathrm{d}v_\tau \\ \qquad + \int\limits_{\partial\mathcal{P}_\tau} \Big(\boldsymbol{t}_{\mathcal{S}}\cdot\boldsymbol{v} + {}^2\boldsymbol{t}_{\mathcal{S}}\cdot\partial_n\boldsymbol{v} + {}^3\boldsymbol{t}_{\mathcal{S}}\cdot\partial_n^2\boldsymbol{v} \\ \qquad + \sum_{\alpha=1}^{n} (\xi^\alpha_{\mathcal{S}}\dot{\varphi}^\alpha + {}^2\xi^\alpha_{\mathcal{S}}\partial_n\dot{\varphi}^\alpha + {}^3\xi^\alpha_{\mathcal{S}}\partial_n^2\dot{\varphi}^\alpha) - \boldsymbol{q}\cdot\boldsymbol{n}\Big)\,\mathrm{d}a_\tau \\ \qquad + \int\limits_{\mathcal{C}_\tau} \Big(\boldsymbol{t}_{\mathcal{C}}\cdot\boldsymbol{v} + {}^2_n\boldsymbol{t}_{\mathcal{C}}\cdot\partial_n\boldsymbol{v} + {}^2_\nu\boldsymbol{t}_{\mathcal{C}}\cdot\partial_\nu\boldsymbol{v} \\ \qquad + \sum_{\alpha=1}^{n} (\xi^\alpha_{\mathcal{C}}\dot{\varphi}^\alpha + {}^2_n\xi^\alpha_{\mathcal{C}}\partial_n\dot{\varphi}^\alpha + {}^2_\nu\xi^\alpha_{\mathcal{C}}\partial_\nu\dot{\varphi}^\alpha)\Big)\,\mathrm{d}\sigma_\tau \\ \qquad + \boldsymbol{t}_{\mathcal{O}}\cdot\boldsymbol{v}|_{\mathcal{O}} + \sum_{\alpha=1}^{n} \xi^\alpha_{\mathcal{O}}\dot{\varphi}^\alpha|_{\mathcal{O}}, \\ \overline{\int\limits_{\mathcal{P}_\tau} \varrho\eta\,\mathrm{d}v_\tau}^{\boldsymbol{\cdot}} \geq \int\limits_{\mathcal{P}_\tau} \frac{r}{\vartheta}\,\mathrm{d}v_\tau - \int\limits_{\partial\mathcal{P}_\tau} \frac{\boldsymbol{q}}{\vartheta}\cdot\boldsymbol{n}\,\mathrm{d}a_\tau. \end{cases} \tag{241}$$

Since $\mathcal{W}_{\mathrm{ext}}(\mathcal{P}_\tau) = \mathcal{W}_{\mathrm{int}}(\mathcal{P}_\tau)$, given by (147), we may substitute the expressions (103) and (105) defining the power expended on $\mathcal{P}_\tau$ by external agencies on the right-hand side of the energy balance $(240)_1$ by the expressions (102) and (104) defining the internal power of $\mathcal{P}_\tau$. In view of the relation (94) arising from the Reynolds' transport theorem, the result of localizing both expressions of (240) is

$$
\begin{cases}
\varrho\dot{\varepsilon} = \mathbf{T} : \operatorname{grad} \boldsymbol{v} + \mathbb{T} \vdots \operatorname{grad}^2 \boldsymbol{v} + \mathbf{T} :: \operatorname{grad}^3 \boldsymbol{v} \\
\qquad + \sum_{\alpha=1}^{n} \Big\{ -\pi^\alpha \dot{\varphi}^\alpha + \boldsymbol{\xi}^\alpha \cdot \operatorname{grad} \dot{\varphi}^\alpha + \boldsymbol{\Xi}^\alpha : \operatorname{grad}^2 \dot{\varphi}^\alpha + \mathbb{X}^\alpha \vdots \operatorname{grad}^3 \dot{\varphi}^\alpha \Big\} \\
\qquad - \operatorname{div} \boldsymbol{q} + r, \\
\varrho\dot{\eta} \geq -\operatorname{div} \dfrac{\boldsymbol{q}}{\vartheta} + \dfrac{r}{\vartheta}.
\end{cases}
\tag{242}
$$

Important to what follows is the free-energy density

$$
\psi := \varepsilon - \vartheta\eta. \tag{243}
$$

Then, since $(242)_2$ may be written as

$$
\varrho\dot{\eta} \geq -\frac{1}{\vartheta} \operatorname{div} \boldsymbol{q} + \frac{1}{\vartheta^2} \boldsymbol{q} \cdot \operatorname{grad} \vartheta + \frac{r}{\vartheta}, \tag{244}
$$

by multiplying this equation by ϑ and subtracting it from $(242)_1$, we arrive at the pointwise free-energy imbalance

$$
\begin{aligned}
&\varrho\dot{\psi} + \varrho\eta\dot{\vartheta} - \mathbf{T} : \operatorname{grad} \boldsymbol{v} - \mathbb{T} \vdots \operatorname{grad}^2 \boldsymbol{v} - \mathbf{T} :: \operatorname{grad}^3 \boldsymbol{v} \\
&+ \sum_{\alpha=1}^{n} \Big\{ \pi^\alpha \dot{\varphi}^\alpha - \boldsymbol{\xi}^\alpha \cdot \operatorname{grad} \dot{\varphi}^\alpha - \boldsymbol{\Xi}^\alpha : \operatorname{grad}^2 \dot{\varphi}^\alpha - \mathbb{X}^\alpha \vdots \operatorname{grad} \dot{\varphi}^\alpha \Big\} \\
&\qquad\qquad + \frac{1}{\vartheta} \boldsymbol{q} \cdot \operatorname{grad} \vartheta =: -\delta \leq 0,
\end{aligned}
\tag{245}
$$

where δ is the entropy production density.

2 Partwise and Pointwise Free-Energy Imbalances for Isothermal Processes for a Spatial Part $\mathcal{P}_\tau$

Applications in which thermal changes are negligible are encompassed by the present framework when attention is restricted to isothermal processes, that is, processes in which

$$
\vartheta = \vartheta_0 = \text{constant}. \tag{246}
$$

For such processes, the free-energy density (243) specializes to

$$
\psi = \varepsilon - \vartheta_0 \eta \tag{247}
$$

and the pointwise free-energy imbalance (245) has a simpler form

$$\varrho\dot{\psi} - \mathbf{T} : \operatorname{grad} \boldsymbol{v} - \mathbb{T} \,⋮\, \operatorname{grad}^2 \boldsymbol{v} - \mathbf{T} :: \operatorname{grad}^3 \boldsymbol{v}$$
$$+ \sum_{\alpha=1}^{n} \left\{ \pi^\alpha \dot{\varphi}^\alpha - \boldsymbol{\xi}^\alpha \cdot \operatorname{grad} \dot{\varphi}^\alpha - \boldsymbol{\Xi}^\alpha : \operatorname{grad}^2 \dot{\varphi}^\alpha - \mathbb{X}^\alpha \,⋮\, \operatorname{grad} \dot{\varphi}^\alpha \right\} =: -\delta \leq 0. \tag{248}$$

Further, multiplying $(242)_2$ by ϑ_0 and subtracting it from $(242)_1$ with (94), we arrive at the partwise free-energy imbalance

$$\overline{\int_{\mathcal{P}_\tau} \varrho\psi \, \mathrm{d}v_\tau}^{\,\boldsymbol{\cdot}} - \mathcal{W}_{\text{ext}}(\mathcal{P}_\tau) =: - \int_{\mathcal{P}_\tau} \delta \, \mathrm{d}v_\tau \leq 0. \tag{249}$$

The expression (249) requires that the temporal increase in free-energy of $\mathcal{P}_\tau$ be less than or equal to the power expended by external agencies on $\mathcal{P}_\tau$. Here, the density $\delta \geq 0$ represents the dissipation per unit volume.

In purely mechanical theories, the imbalance (249) is often the implicit point of departure. Naturally, the pointwise imbalance (248) may be derived directly, without introducing the notion of temperature, from (249), the power balances (100) and (101), and the expressions (102) and (104) for the internal power. In that sense, the imbalance stands on its own as a point of departure for the development of purely mechanical theories.

Next, with the material derivative given as $\dot{\varphi} = \partial_t \varphi + \boldsymbol{v} \cdot \operatorname{grad} \varphi$, consider the following identities related to commutating the material time derivative with the spatial first, second, and third gradients.

$$\operatorname{grad} \dot{\varphi}^\alpha = (\operatorname{grad} \varphi^\alpha)^{\boldsymbol{\cdot}} + (\operatorname{grad} \boldsymbol{v})^{\top}[\operatorname{grad} \varphi^\alpha], \tag{250}$$

$$\operatorname{grad}^2 \dot{\varphi}^\alpha = (\operatorname{grad}^2 \varphi^\alpha)^{\boldsymbol{\cdot}} + (\operatorname{grad}^2 \boldsymbol{v})^{\vdash}[\operatorname{grad} \varphi^\alpha] + 2(\operatorname{grad} \boldsymbol{v})^{\top} \operatorname{grad}^2 \varphi^\alpha, \tag{251}$$

and

$$\operatorname{grad}^3 \dot{\varphi}^\alpha = (\operatorname{grad}^3 \varphi^\alpha)^{\boldsymbol{\cdot}} + (\operatorname{grad}^3 \boldsymbol{v})^{\vdash}[\operatorname{grad} \varphi^\alpha] + 3(\operatorname{grad}^2 \boldsymbol{v})^{\vdash} \operatorname{grad}^2 \varphi^\alpha$$
$$+ 3(\operatorname{grad} \boldsymbol{v})^{\top} \cdot \operatorname{grad}^3 \varphi^\alpha. \tag{252}$$

With (250)–(252), the pointwise free-energy imbalance (248) assumes the form

$$
\begin{aligned}
0 \geq \varrho\dot{\psi} &- (\mathbf{T} + \sum_{\alpha=1}^{n}(\operatorname{grad}\varphi^\alpha \otimes \boldsymbol{\xi}^\alpha + 2(\operatorname{grad}^2\varphi^\alpha)\boldsymbol{\Xi}^\alpha + 3\operatorname{grad}^3\varphi^\alpha : \mathbb{X}^\alpha)) : \operatorname{grad}\boldsymbol{v} \\
&- (\mathbb{T} + \sum_{\alpha=1}^{n}(\operatorname{grad}\varphi^\alpha \otimes \boldsymbol{\Xi}^\alpha + 3\operatorname{grad}^2\varphi^\alpha \cdot \mathbb{X}^\alpha)) \,\vdots\, \operatorname{grad}^2\boldsymbol{v} \\
&- (\mathbb{T} + \sum_{\alpha=1}^{n}\operatorname{grad}\varphi^\alpha \otimes \mathbb{X}^\alpha) :: \operatorname{grad}^3\boldsymbol{v} \\
&+ \sum_{\alpha=1}^{n}\Big\{\pi^\alpha\dot{\varphi}^\alpha - \boldsymbol{\xi}^\alpha \cdot (\operatorname{grad}\varphi^\alpha)^{\cdot} - \boldsymbol{\Xi}^\alpha : (\operatorname{grad}^2\varphi^\alpha)^{\cdot} - \mathbb{X}^\alpha \,\vdots\, (\operatorname{grad}^3\varphi^\alpha)^{\cdot}\Big\}.
\end{aligned}
\tag{253}
$$

3 First and Second Laws of Thermodynamics for a Control Volume $\overline{\mathcal{P}}$

Integrating (242) in a control volume $\overline{\mathcal{P}}$ followed by $(97)_1$, the first and second law of thermodynamics read

$$
\begin{cases}
\overline{\int_{\overline{\mathcal{P}}} \varrho(\varepsilon + \tfrac{1}{2}|\boldsymbol{v}|^2)\,\mathrm{d}v}^{\,\cdot} + \\
\int_{\partial\overline{\mathcal{P}}} (\varrho\varepsilon + \tfrac{1}{2}|\boldsymbol{v}|^2)\boldsymbol{v}\cdot\boldsymbol{n}\,\mathrm{d}v = \mathcal{W}_{\text{ext}}^{\text{conv}}(\overline{\mathcal{P}}) + \int_{\overline{\mathcal{P}}} r\,\mathrm{d}v - \int_{\partial\overline{\mathcal{P}}} \boldsymbol{q}\cdot\boldsymbol{n}\,\mathrm{d}a, \\
\overline{\int_{\overline{\mathcal{P}}} \varrho\eta\,\mathrm{d}v}^{\,\cdot} + \int_{\partial\overline{\mathcal{P}}} \varrho\eta\,\boldsymbol{v}\cdot\boldsymbol{n}\,\mathrm{d}v \geq \int_{\overline{\mathcal{P}}} \frac{r}{\vartheta}\,\mathrm{d}v - \int_{\partial\overline{\mathcal{P}}} \frac{\boldsymbol{q}}{\vartheta}\cdot\boldsymbol{n}\,\mathrm{d}a.
\end{cases}
\tag{254}
$$

4 Partwise Free-Energy Imbalances for Isothermal Processes for a Control Volume $\overline{\mathcal{P}}$

Emulating the process that led us to (249) from (240), we depart from (254) to arrive at

$$
\overline{\int_{\overline{\mathcal{P}}} \varrho\psi\,\mathrm{d}v}^{\,\cdot} + \int_{\partial\overline{\mathcal{P}}} \varrho\psi\,\boldsymbol{v}\cdot\boldsymbol{n}\,\mathrm{d}v - \mathcal{W}_{\text{ext}}(\overline{\mathcal{P}}) =: -\int_{\overline{\mathcal{P}}} \delta\,\mathrm{d}v \leq 0. \tag{255}
$$

5 Derivatives, Inner Products, and Consequences in Constrained Spaces

Here, we employ ideas by Larché & Cahn [54] and Fried & Gurtin [55] to deal with derivatives with respect to the phase fields when constrained by an additional equality condition. In a system of n phase fields, let $n \in \mathbb{N}$ and φ^α be the phase field representing the αth component and the set of phase fields be represented by

$$\boldsymbol{\varphi} = \{\varphi^1, \ldots, \varphi^n\}. \tag{256}$$

Regardless of the underlying physical description of the phase fields, in various physical settings the following constraint holds

$$\sum_{\alpha=1}^{n} \varphi^\alpha = 1. \tag{257}$$

A new set of constraints may be derived by computing the mth gradient of the constraint (257), that is,

$$\sum_{\alpha=1}^{n} \operatorname{grad}^m \varphi^\alpha = \mathbf{0}. \tag{258}$$

Within this constrained setting, let $\tilde{\boldsymbol{\varphi}} = \boldsymbol{\varphi} \setminus \{\varphi^\varsigma\}$ be the reduced set of $(n-1)$-independent phase-field components, where the missing component φ^ς may be recovered from $\tilde{\boldsymbol{\varphi}}$, that is,

$$\varphi^\varsigma = 1 - \sum_{\substack{\alpha=1 \\ \alpha \neq \varsigma}}^{n} \varphi^\alpha, \tag{259}$$

and, analogously, the mth gradient of the missing component is

$$\operatorname{grad}^m \varphi^\varsigma = -\sum_{\substack{\alpha=1 \\ \alpha \neq \varsigma}}^{n} \operatorname{grad}^m \varphi^\alpha. \tag{260}$$

5.1 *Derivatives*

Now, let the functional $\mathcal{L}^0$ depend on $\boldsymbol{\varphi}$ such that

$$\mathcal{L}^0(\boldsymbol{\varphi}) := \mathcal{L}^0(\varphi^1, \ldots, \varphi^n). \tag{261}$$

Since a variation on the αth phase field φ^α when holding the others fixed violates the constraint (257) and, consequently, (258), conventional partial derivatives are not well defined. Thus, we need to define a differential operator such that (257) holds. An important aspect of this theory is that the mappings (259) and (260) must be taken into account when computing partial derivatives with respect to φ^α and $\mathrm{grad}^m \varphi^\alpha$. Larché & Cahn [54] devised a derivative operation to overcome this issue.

Definition 1 *(Larché–Cahn derivative)* In a constrained setting of $n \in \mathbb{N}$ phase fields with $n-1$ independent components, the Larché–Cahn derivative of the functional $\mathcal{L}^0$ given in expression (261) with respect to the αth phase-field component, satisfying the constraint (257), is given by

$$\frac{\partial^{(\varsigma)} \mathcal{L}^0(\boldsymbol{\varphi})}{\partial \varphi^\alpha} := \frac{\mathrm{d}}{\mathrm{d}\epsilon} \mathcal{L}^0(\varphi^1, \ldots, \varphi^\alpha + \epsilon, \ldots, \varphi^\varsigma - \epsilon, \ldots, \varphi^n)\bigg|_{\epsilon=0}, \tag{262}$$

where the αth and ςth phase-field components suffer opposite variations. Following the same reasoning, second-order Larché–Cahn derivatives are defined as follows.

$$\frac{\partial^{2(\varsigma)} \mathcal{L}^0(\boldsymbol{\varphi})}{\partial \varphi^\alpha \, \partial \varphi^\beta} := \frac{\mathrm{d}}{\mathrm{d}\epsilon_1 \, \mathrm{d}\epsilon_2} \mathcal{L}^0(\varphi^1, \ldots, \varphi^\alpha + \epsilon_1, \ldots, \varphi^\beta + \epsilon_2, \ldots, \varphi^\varsigma - \epsilon_1 - \epsilon_2, \ldots, \varphi^n)\bigg|_{\substack{\epsilon_1=0 \\ \epsilon_2=0}}. \tag{263}$$

In the Larché–Cahn derivative (262), to compute a variation with respect to the phase field φ^α, the phase field φ^ς is chosen as a reference component in the set of variables. An infinitesimal change, $+\epsilon$, in φ^α induces the opposite infinitesimal variation, $-\epsilon$, into φ^ς when holding the others unchanged.

Property 1 (Derivative of the reference and skew-symmetric relation arising from Definition 1) Immediate consequences of the Larché–Cahn derivative (262) of Definition (1) are given by

$$\frac{\partial^{(\varsigma)} \mathcal{L}^0(\boldsymbol{\varphi})}{\partial \varphi^\varsigma} = 0, \tag{264}$$

and the skew-symmetric relation

$$\frac{\partial^{(\varsigma)} \mathcal{L}^0(\boldsymbol{\varphi})}{\partial \varphi^\alpha} = -\frac{\partial^{(\alpha)} \mathcal{L}^0(\boldsymbol{\varphi})}{\partial \varphi^\varsigma}. \tag{265}$$

Property 2 (Degeneration of the Larché–Cahn derivative) Let $0 \le \varphi^\alpha \le 1$ without loss of generality. In the limiting case, where the reference component φ^ς equals zero, or if the component of interest φ^α equals one, at some $\boldsymbol{x}$ and t, the Larché–Cahn derivative is also not well defined since $\varphi^\varsigma - \epsilon < 0$, or $\varphi^\alpha + \epsilon > 1$, at some $\boldsymbol{x}$ and t. However, owing to the Property (265), we redefine the Larché–Cahn derivative for this particular case as

$$\frac{\partial^{(\varsigma)} \mathcal{L}^0(\boldsymbol{\varphi})}{\partial \varphi^\alpha} := -\frac{\partial^{(\alpha)} \mathcal{L}^0(\boldsymbol{\varphi})}{\partial \varphi^\varsigma}. \tag{266}$$

In view of the Property 2, one could instead allow a smooth extension of $\{\varphi^\alpha\}_{\alpha=1}^n$ to an open region of $\mathbb{R}^n$.

Next, emulating the reasoning that led us to define the Larché–Cahn derivative (262), we consider functionals depending upon the gradients of the phase fields. Thus, let the functional $\mathcal{L}^m$ depend on $\operatorname{grad}^m \boldsymbol{\varphi}$ such that

$$\mathcal{L}^m(\operatorname{grad}^m \boldsymbol{\varphi}) := \mathcal{L}^m(\operatorname{grad}^m \varphi^1, \ldots, \operatorname{grad}^m \varphi^n). \tag{267}$$

We are now led to define an extended version of the Larché–Cahn derivative when mth gradients of the phase field enter as arguments of a functional $\mathcal{L}^m$ as follows.

Definition 2 *(Extended Larché–Cahn derivative)* In a constrained setting of $n \in \mathbb{N}$ phase fields with $n-1$ independent components, the extended Larché–Cahn derivative of a functional $\mathcal{L}^m$ given in (267) with respect to the mth gradient of the αth phase-field component, satisfying the constraint (258), is given by

$$\frac{\partial^{(\varsigma)} \mathcal{L}^m(\operatorname{grad}^m \boldsymbol{\varphi})}{\partial(\operatorname{grad}^m \varphi^\alpha)} := \frac{\mathrm{d}}{\mathrm{d}\epsilon}\mathcal{L}^m(\operatorname{grad}^m \varphi^1, \ldots, \operatorname{grad}^m \varphi^\alpha + \epsilon \boldsymbol{o}^{(m)}, \ldots, \operatorname{grad}^m \varphi^\varsigma - \epsilon \boldsymbol{o}^{(m)}, \ldots, \operatorname{grad}^m \varphi^n)\Big|_{\epsilon=0}, \tag{268}$$

where $\boldsymbol{o}^{(m)}$ is a mth order tensor fully populated with ones. The second-order extended Larché–Cahn derivative reads

$$\frac{\partial^{2(\varsigma)} \mathcal{L}^m(\operatorname{grad}^m \boldsymbol{\varphi})}{\partial(\operatorname{grad}^m \varphi^\alpha)\,\partial(\operatorname{grad}^m \varphi^\beta)} := \frac{\mathrm{d}}{\mathrm{d}\epsilon_1\,\mathrm{d}\epsilon_2}\mathcal{L}^m(\operatorname{grad}^m \varphi^1, \ldots, \operatorname{grad}^m \varphi^\alpha + \epsilon_1 \boldsymbol{o}^{(m)}, \ldots, \operatorname{grad}^m \varphi^\beta + \epsilon_2 \boldsymbol{o}^{(m)}, \ldots, \operatorname{grad}^m \varphi^\varsigma - \epsilon_1 \boldsymbol{o}^{(m)} - \epsilon_2 \boldsymbol{o}^{(m)}, \ldots, \operatorname{grad}^m \varphi^n)\Big|_{\substack{\epsilon_1=0\\ \epsilon_2=0}}. \tag{269}$$

Property 3 (Derivative of the reference and skew-symmetric relation arising from Definition 2) Analogously to Properties 1, immediate consequences arising from expression (268) in Definition 2 are given by

$$\frac{\partial^{(\varsigma)} \mathcal{L}^m(\operatorname{grad}^m \boldsymbol{\varphi})}{\partial(\operatorname{grad}^m \varphi^\varsigma)} = 0, \tag{270}$$

and the skew-symmetric relation

$$\frac{\partial^{(\varsigma)} \mathcal{L}^m(\operatorname{grad}^m \boldsymbol{\varphi})}{\partial(\operatorname{grad}^m \varphi^\alpha)} = -\frac{\partial^{(\alpha)} \mathcal{L}^m(\operatorname{grad}^m \boldsymbol{\varphi})}{\partial(\operatorname{grad}^m \varphi^\varsigma)}. \tag{271}$$

Naturally, Property 2 does not have a reciprocal version in considering the functional $\mathcal{L}^m$. Thus, no redefinition of the extended Larché–Cahn derivative is needed.

Finally, with a smooth extension of $\mathcal{L}^0$ and $\mathcal{L}^m$ to an open region of $\mathbb{R}^n$, the Larché–Cahn derivatives (262) and (268) can be computed as follows.

$$\frac{\partial^{(\varsigma)}\mathcal{L}^0(\boldsymbol{\varphi})}{\partial\varphi^\alpha} = \frac{\partial\mathcal{L}^0(\boldsymbol{\varphi})}{\partial\varphi^\alpha} - \frac{\partial\mathcal{L}^0(\boldsymbol{\varphi})}{\partial\varphi^\varsigma}, \tag{272}$$

and

$$\frac{\partial^{(\varsigma)}\mathcal{L}^m(\operatorname{grad}^m\boldsymbol{\varphi})}{\partial(\operatorname{grad}^m\varphi^\alpha)} = \frac{\partial\mathcal{L}^m(\operatorname{grad}^m\boldsymbol{\varphi})}{\partial(\operatorname{grad}^m\varphi^\alpha)} - \frac{\partial\mathcal{L}^m(\operatorname{grad}^m\boldsymbol{\varphi})}{\partial(\operatorname{grad}^m\varphi^\varsigma)}. \tag{273}$$

5.2 Inner Products

In a system of multicomponent along with the free-energy density definition, we often encounter inner products in the following form

$$\sum_{\alpha=1}^{n}\sum_{\beta=1}^{n}\operatorname{grad}^m\varphi^\alpha \cdot \boldsymbol{\Gamma}^{\alpha\beta}[\operatorname{grad}^m\varphi^\beta], \tag{274}$$

where $\boldsymbol{\Gamma}$ is a tensor of order $2m$. However, for the point we wish to make, without loss of generality, let $m = 1$.

Consider a set of vectors $\{\boldsymbol{p}^\alpha\}_{\alpha=1}^n := \{\operatorname{grad}^m\varphi^\alpha\}_{\alpha=1}^n$ constrained by

$$\sum_{\alpha=1}^{n}\boldsymbol{p}^\alpha = \mathbf{0}, \tag{275}$$

with the inner product (274) rewritten as

$$\sum_{\alpha=1}^{n}\sum_{\beta=1}^{n}\boldsymbol{p}^\alpha \cdot \boldsymbol{\Gamma}^{\alpha\beta}[\boldsymbol{p}^\beta]. \tag{276}$$

Property 4 (Nontrivial nullspaces in constrained spaces) Let $\boldsymbol{\Theta}^{\alpha\beta}$ be populated with $\boldsymbol{\Theta}$ for all α and β from 1 to n, that is,

$$\boldsymbol{\Theta}^{\alpha\beta} := \boldsymbol{\Theta} \qquad \forall\, 1 \leq \alpha, \beta \leq n. \tag{277}$$

Thus, the set $\{\boldsymbol{p}^\alpha\}_{\alpha=1}^n$ satisfying the constraint (275) is the nullspace of $\boldsymbol{\Theta}^{\alpha\beta}$, that is, $\operatorname{Null}(\boldsymbol{\Theta}^{\alpha\beta}) = \{\boldsymbol{p}^\alpha\}_{\alpha=1}^n$. Therefore, from expression (276), we have that

$$\sum_{\alpha=1}^{n}\sum_{\beta=1}^{n} \boldsymbol{p}^{\alpha} \cdot \boldsymbol{\Gamma}^{\alpha\beta}[\boldsymbol{p}^{\beta}] = \sum_{\alpha=1}^{n}\sum_{\beta=1}^{n} \boldsymbol{p}^{\alpha} \cdot (\boldsymbol{\Gamma}^{\alpha\beta} + \boldsymbol{\Theta}^{\alpha\beta})[\boldsymbol{p}^{\beta}]. \tag{278}$$

Next, imposing the constraint (275) on the component ς in the inner product (276), we arrive at

$$\sum_{\alpha=1}^{n}\sum_{\beta=1}^{n} \boldsymbol{p}^{\alpha} \cdot \boldsymbol{\Gamma}^{\alpha\beta}[\boldsymbol{p}^{\beta}] = \sum_{\substack{\alpha=1 \\ \alpha\neq\varsigma}}^{n}\sum_{\substack{\beta=1 \\ \beta\neq\varsigma}}^{n} \boldsymbol{p}^{\alpha} \cdot \underbrace{(\boldsymbol{\Gamma}^{\alpha\beta} + \boldsymbol{\Gamma}^{\varsigma\varsigma} - \boldsymbol{\Gamma}^{\alpha\varsigma} - \boldsymbol{\Gamma}^{\varsigma\beta})}_{\boldsymbol{\Gamma}^{\alpha\beta}_{\varsigma}}[\boldsymbol{p}^{\beta}]. \tag{279}$$

Thus, we are led to define the effective metrics in constrained spaces as follows.

Definition 3 *(Effective metrics in constrained spaces)* Via *non-invertible mappings* $\boldsymbol{\Gamma}^{\alpha\beta} \mapsto \boldsymbol{\Gamma}^{\alpha\beta}_{\varsigma}$, the metric $\boldsymbol{\Gamma}^{\alpha\beta}_{\varsigma}$ of the vector space defined by the elements satisfying the constraint (275) is defined by

$$\boldsymbol{\Gamma}^{\alpha\beta}_{\varsigma} := \boldsymbol{\Gamma}^{\alpha\beta} + \boldsymbol{\Gamma}^{\varsigma\varsigma} - \boldsymbol{\Gamma}^{\alpha\varsigma} - \boldsymbol{\Gamma}^{\varsigma\beta}. \tag{280}$$

Furthermore, the inner product in (274) does not depend upon the choice of the reference phase field φ^{ς}.

6 Representation Theorem for Isotropic Tensors

What we require first is the concept of an irreducible representation of a group. To that end, we use the proper orthogonal group SO(3) of three-dimensional rotations as the standard form. Let V be a vector space and the space of invertible linear operators (endomorphisms) acting on V be $\mathrm{GL}(V)$. We can then represent the group on V via a mapping θ: SO(3) $\rightarrow$ $\mathrm{GL}(V)$, meaning that for $\mathbf{Q} \in \mathrm{SO}(3)$ and $\boldsymbol{v} \in V$ we have the rotated vector $\theta(\mathbf{Q})[\boldsymbol{v}] \in V$. Thus, we can say that V is a 'representation' of the group since its elements are related by a group action.

There are certain special 'building block' representations known as irreducible representations that can be determined by looking at invariant subspaces. Suppose $U \subset V$. If for $\boldsymbol{u} \in U$ we have that $\theta(\mathbf{Q})[\boldsymbol{u}] \in U$ for all $\mathbf{Q} \in \mathrm{SO}(3)$, then V is known as a *reducible* representation if U is a proper subspace. Conversely, if V is a *irreducible* representation, all such U are trivial, that is, either V itself or the nullspace. Basically, if V is a representation and it contains nontrivial invariant subspaces, it is reducible. Then, we can further check if these invariant subspaces contain further invariant subspaces. Eventually, we reach a point where we cannot go any further and are left with trivial subspaces. Thus, we have broken down the original reducible representation into irreducible ones. All the irreducible representations of a group can be determined directly from the group and are intimately connected to the group itself.

For instance, let V by a reducible representation, and let it contain invariant subspaces U_1 and U_2, that is, $V = U_1 \oplus U_2$. Analogously, let U_1 and U_2 contain, respectively, the additional invariant subspaces U_{11}, U_{12} and U_{21}, U_{22}. Thus, we have $V = (U_{11} \oplus U_{12}) \oplus (U_{21} \oplus U_{22})$. Next, suppose the U_{ij} do not have any more proper invariant subspaces, only themselves and the nullspace. Then, they are irreducible, and we have decomposed the reducible representation V into a sum of irreducible representations.

6.1 Decomposition of Tensors

We label the irreducible representation of SO(3) as Γ_j where the dimension of the representations (the size of the matrix that acts on the given vector space) is $2j + 1$. So Γ_0 is one-dimensional and is the trivial representation (it does not change anything), Γ_1 is three-dimensional (it acts on three-component vectors), Γ_2 is five-dimensional, and so on. How to apply this to the tensors at hand? We want to build tensors that transform in these irreducible representations (for instance, if it is in Γ_0, the trivial representation, then it does not change under rotations, that is, it is isotropic, et cetera). Let us say V is a three-dimensional vector space. Then a first-order tensor with components v^i where $i \in \{1, 2, 3\}$ is a vector and Γ_1 is a three-dimensional irreducible representation, so we have v^i transforming in Γ_1.

Now, let C_{ij} be the components of a second-order tensor on a three-dimensional vector space. Tensors are built from the tensor product of vectors and each index transforms in Γ_1, so the second-order tensor will be in $\Gamma_1 \otimes \Gamma_1$, which is the tensor product of the three-dimensional representations and so is nine-dimensional, indeed, a second-order tensor has nine-independent components. This tensor product is reducible and the Clebsch–Gordan decomposition theorem allows us to decompose the tensor products of irreducible representations of SO(3) (and other groups too) into the sum of irreducible representations as

$$\Gamma_i \otimes \Gamma_j = \bigoplus_{k=|i-j|}^{i+j} \Gamma_k = \Gamma_{|i-j|} \oplus \Gamma_{|i-j|+1} \oplus \cdots \oplus \Gamma_{i+j}, \tag{281}$$

For instance, consider $\Gamma_1 \otimes \Gamma_1 = \Gamma_0 \oplus \Gamma_1 \oplus \Gamma_2$. Thus, we can decompose a second-order tensor into three terms, one that is in Γ_0, that is, invariant under rotations and corresponds to one degree of freedom, one that is in Γ_1 and transforms as a three-dimensional object and thus takes three degrees of freedom, and one term in Γ_2 that is five-dimensional and thus takes five degrees of freedom, in total 1+3+5=9. So all 9 original independent components (or degrees of freedom) are accounted for. How to do this concretely? Consider the following second-order tensor

$$C_{ij} = \tfrac{1}{2}(C_{ij} + C_{ji}) + \tfrac{1}{2}(C_{ij} - C_{ji}),$$
$$= \underbrace{\tfrac{1}{3}C_{kk}\delta_{ij}}_{\Gamma_0} + \underbrace{\tfrac{1}{2}(C_{ij} - C_{ji})}_{\Gamma_1} + \underbrace{\tfrac{1}{2}(C_{ij} + C_{ji} - \tfrac{2}{3}C_{kk}\delta_{ij})}_{\Gamma_2}, \tag{282}$$

or

$$\mathbf{C} = \tfrac{1}{3}\mathrm{tr}\,(\mathbf{C})\mathbf{1} + \tfrac{1}{2}(\mathbf{C} - \mathbf{C}^{\top}) + \tfrac{1}{2}(\mathbf{C} + \mathbf{C}^{\top} - \tfrac{2}{3}\mathrm{tr}\,(\mathbf{C})\mathbf{1}). \tag{283}$$

Indeed, the trace part is invariant under rotations, the antisymmetric part has three independent components, and the traceless symmetric part has five independent components.

Moreover, this decomposition shows a connection between the permutations of the tensor indices and the irreducible representations. This connection has a detailed theory for other groups as well and can be found from so-called Young-diagrams, or Young-tableaux. The interested reader is referred to the book by Dresselhaus et al. [56].

As mentioned, when it comes to analyzing invariant components under SO(3), only the copies of Γ_0 are relevant. Thus, representation (282) is reduced to

$$C_{ij} = \tfrac{1}{3}C_{kk}\delta_{ij},$$
$$=: \ell_{(2,1)}\delta_{ij}, \tag{284}$$

or

$$\mathbf{C} = \tfrac{1}{3}\mathrm{tr}\,(\mathbf{C})\mathbf{1},$$
$$=: \ell_{(2,1)}\mathbf{1}, \tag{285}$$

where $\ell_{(\iota,b)}$, in what follows, represents the bth scalar coefficient for a tensor of order ι. Next, for simplicity, we will be using 'components of a tensor' instead of the 'tensor', but only in this part of this study.

6.2 Fourth-Order Isotropic Tensor

Consider the fourth-order tensor C_{ijkl} where for the general case we have

$$\bigotimes_{i=1}^{4}\Gamma_1 = \Gamma_1 \otimes \Gamma_1 \otimes \Gamma_1 \otimes \Gamma_1,$$
$$= (\Gamma_0 \oplus \Gamma_1 \oplus \Gamma_2) \otimes (\Gamma_0 \oplus \Gamma_1 \oplus \Gamma_2),$$
$$= 3\Gamma_0 \oplus 6\Gamma_1 \oplus 6\Gamma_2 \oplus 3\Gamma_3 \oplus \Gamma_4. \tag{286}$$

There are three copies of the trivial irrep (irreducible representations) Γ_0, so, in general, we require three coefficients $\ell_{(4,b)}$, with $1 \leq b \leq 3$, to describe a fourth-order isotropic tensor. The terms in it contain independent combinations of two Kronecker deltas, that is,

$$(\delta_{ij}\delta_{kl};\ \delta_{ik}\delta_{jl};\ \delta_{il}\delta_{jk}). \tag{287}$$

Assuming certain symmetries, these may reduce the number of independent terms. First, consider the major symmetry needed to impose the symmetry of the inner product: $(ij) \leftrightarrow (kl)$. All three possible terms in (287) already satisfy this symmetry. Thus, this requirement does not reduce the number of terms.

Next, we consider the minor symmetry: $i \leftrightarrow j$. Symmetrizing the last two terms in (287), we find that these are no longer independent, thus

$$(\delta_{ij}\delta_{kl};\ \delta_{ik}\delta_{jl};\ \delta_{il}\delta_{jk}) \to (\delta_{ij}\delta_{kl};\ \delta_{ik}\delta_{jl} + \delta_{il}\delta_{jk}). \tag{288}$$

The result of enforcing the minor symmetries is that the number of independent terms is reduced by one, and we are left with two independent terms, so the most general fourth-order isotropic tensor is

$$C^{\Gamma_0}_{ijkl} = \ell_{(4,1)}\delta_{ij}\delta_{kl} + \ell_{(4,2)}(\delta_{ik}\delta_{jl} + \delta_{il}\delta_{jk}). \tag{289}$$

Contracting with a symmetric second-order tensor A_{ij}, we arrive at

$$C^{\Gamma_0}_{ijkl}A_{kl} = \ell_{(4,1)}\delta_{ij}A_{kk} + 2\ell_{(4,2)}A_{ij}, \tag{290}$$

giving the usual form. Also, note that if we do not require the minor symmetry (but contract it with a nonsymmetric second-order tensor), then we need three independent coefficients to describe the tensor, with the third one corresponding to the antisymmetric part. Furthermore, the norm squared of A_{ij} reads

$$\begin{aligned} \|\mathbf{A}\|^2 &:= A_{ij}C^{\Gamma_0}_{ijkl}A_{kl}, \\ &= \ell_{(4,1)}A_{ii}A_{kk} + 2\ell_{(4,2)}A_{ij}A_{ij}. \end{aligned} \tag{291}$$

If we further specialize these relations for the case of a theory involving incompressible materials, where A_{ij} satisfies $\mathbf{1} : \mathbf{A} = 0$, (290) and (291) become

$$C^{\Gamma_0}_{ijkl}A_{kl} = 2\ell_{(4,2)}A_{ij}, \tag{292}$$

and

$$\begin{aligned} \|\mathbf{A}\|^2 &:= A_{ij}C^{\Gamma_0}_{ijkl}A_{kl}, \\ &= 2\ell_{(4,2)}A_{ij}A_{ij}. \end{aligned} \tag{293}$$

6.3 Sixth-Order Isotropic Tensor

Consider the sixth-order tensor C_{ijklmn} where for the general case we have

$$\begin{aligned}\bigotimes_{i=1}^{6}\Gamma_1 &= \Gamma_1\otimes\Gamma_1\otimes\Gamma_1\otimes\Gamma_1\otimes\Gamma_1\otimes\Gamma_1,\\ &= (\Gamma_0\oplus\Gamma_1\oplus\Gamma_2)\otimes(\Gamma_0\oplus\Gamma_1\oplus\Gamma_2)\otimes(\Gamma_0\oplus\Gamma_1\oplus\Gamma_2),\\ &= (3\Gamma_0\oplus 6\Gamma_1\oplus 6\Gamma_2\oplus 3\Gamma_3\oplus\Gamma_4)\otimes(\Gamma_0\oplus\Gamma_1\oplus\Gamma_2),\\ &= 15\Gamma_0\oplus 36\Gamma_1\oplus 40\Gamma_2\oplus 29\Gamma_3\oplus 15\Gamma_4\oplus 5\Gamma_5\oplus\Gamma_6.\end{aligned}\tag{294}$$

We see 15 copies of the trivial irrep Γ_0, so we need 15 coefficients to describe the most general sixth-order isotropic tensor. The 15 possible terms are independent combinations of three Kronecker deltas

$$\begin{aligned}(&\delta_{ij}\delta_{kl}\delta_{mn};\ \delta_{ij}\delta_{km}\delta_{ln};\ \delta_{ij}\delta_{kn}\delta_{lm};\\ &\delta_{ik}\delta_{jl}\delta_{mn};\ \delta_{ik}\delta_{jm}\delta_{ln};\ \delta_{ik}\delta_{jn}\delta_{lm};\\ &\delta_{il}\delta_{jk}\delta_{mn};\ \delta_{il}\delta_{jm}\delta_{kn};\ \delta_{il}\delta_{jn}\delta_{km};\\ &\delta_{im}\delta_{jk}\delta_{ln};\ \delta_{im}\delta_{jl}\delta_{kn};\ \delta_{im}\delta_{jn}\delta_{kl};\\ &\delta_{in}\delta_{jk}\delta_{lm};\ \delta_{in}\delta_{jm}\delta_{kl};\ \delta_{in}\delta_{jl}\delta_{km}).\end{aligned}\tag{295}$$

Next, consider the major symmetry needed to impose the symmetry of the inner product $(ijk)\leftrightarrow(lmn)$. In order to enforce this symmetry, we have to combine some of these terms, thereby reducing the number of independent terms from 15 to 11. The minor symmetry $j\leftrightarrow k$, combined with the major symmetry, implies another symmetry, namely, $m\leftrightarrow n$. With these minor symmetries, we have five remaining terms, thus, we need five coefficients $\ell_{(6,b)}$, with $1\le b\le 5$, to describe a sixth-order isotropic tensor with the given symmetries. Altogether, we see how the symmetries reduce the number of independent terms from 15 to 5. Indeed, from (295) while imposing the major symmetry and the mentioned minor symmetries, we have that

$$\begin{aligned}C^{\Gamma_0}_{ijklmn} &= \ell_{(6,1)}(\delta_{ij}\delta_{kl}\delta_{mn}+\delta_{ik}\delta_{jl}\delta_{mn}+\delta_{in}\delta_{jk}\delta_{lm}+\delta_{im}\delta_{jk}\delta_{ln})\\ &+\ell_{(6,2)}(\delta_{ij}\delta_{km}\delta_{ln}+\delta_{ik}\delta_{jm}\delta_{ln}+\delta_{ij}\delta_{kn}\delta_{lm}+\delta_{ik}\delta_{jn}\delta_{lm})\\ &+\ell_{(6,3)}(\delta_{im}\delta_{jn}\delta_{kl}+\delta_{im}\delta_{jl}\delta_{kn}+\delta_{in}\delta_{jm}\delta_{kl}+\delta_{in}\delta_{jl}\delta_{km})\\ &+\ell_{(6,4)}(\delta_{il}\delta_{jm}\delta_{kn}+\delta_{il}\delta_{jn}\delta_{km})+\ell_{(6,5)}\delta_{il}\delta_{jk}\delta_{mn}.\end{aligned}\tag{296}$$

Contracting $C^{\Gamma_0}_{ijklmn}$ with a third-order tensor A_{ijk} that satisfies $A_{ijk}=A_{ikj}$, then

$$\begin{aligned} C^{\Gamma_0}_{ijklmn}A_{lmn} = &\ \ell_{(6,1)}(\delta_{ij}A_{kmm} + \delta_{ik}A_{jmm} + 2\delta_{jk}A_{mmi}) \\ &+ 2\ell_{(6,2)}(\delta_{ij}A_{mmk} + \delta_{ik}A_{mmj}) \\ &+ 2\ell_{(6,3)}(A_{kij} + A_{jik}) \\ &+ 2\ell_{(6,4)}A_{ijk} + \ell_{(6,5)}\delta_{jk}A_{imm}. \end{aligned} \tag{297}$$

Furthermore, the norm squared of A_{ijk} reads

$$\begin{aligned} \|\mathbb{A}\|^2 :=& A_{ijk}C^{\Gamma_0}_{ijklmn}A_{lmn}, \\ =&\ 4\ell_{(6,1)}A_{iik}A_{kmm} + 4\ell_{(6,2)}A_{iik}A_{mmk} \\ &+ 2\ell_{(6,3)}(A_{ijk}A_{kij} + A_{ijk}A_{jik}) \\ &+ 2\ell_{(6,4)}A_{ijk}A_{ijk} + \ell_{(6,5)}A_{ikk}A_{imm}. \end{aligned} \tag{298}$$

Once again, for the continuum type of theory involving incompressible materials, consider that $\mathbb{A}$ satisfies $\mathbf{1} : \mathbb{A} = \mathbf{0}$, then, expressions (297) specializes to

$$\begin{aligned} C^{\Gamma_0}_{ijklmn}A_{lmn} = &\ \ell_{(6,1)}(\delta_{ij}A_{kmm} + \delta_{ik}A_{jmm}) \\ &+ 2\ell_{(6,3)}(A_{kij} + A_{jik}) \\ &+ 2\ell_{(6,4)}A_{ijk} + \ell_{(6,5)}\delta_{jk}A_{imm}. \end{aligned} \tag{299}$$

Letting $S_{ijk} := C^{\Gamma_0}_{ijklmn}A_{lmn}$, without loss of generality, we impose $\mathbf{1} : \mathbb{S} = \mathbf{0}$ and from (299), we arrive at the following condition: $4\ell_{(6,1)} + 2\ell_{(6,3)} + \ell_{(6,5)} = 0$. Moreover, given the powerless nature of the term multiplying $\ell_{(6,1)}$ in expression (298), we opt for $\ell_{(6,1)} = -\frac{1}{2}(\ell_{(6,3)} + \frac{1}{2}\ell_{(6,5)})$ and expression (299) is specialized to

$$\begin{aligned} C^{\Gamma_0}_{ijklmn}A_{lmn} = &\ 2\ell_{(6,3)}(A_{kij} + A_{jik} - \tfrac{1}{2}(\delta_{ij}A_{kmm} + \delta_{ik}A_{jmm})) \\ &+ 2\ell_{(6,4)}A_{ijk} + \ell_{(6,5)}(\delta_{jk}A_{imm} - \tfrac{1}{2}(\delta_{ij}A_{kmm} + \delta_{ik}A_{jmm})), \end{aligned} \tag{300}$$

while (298) becomes

$$\begin{aligned} \|\mathbb{A}\|^2 :=&\ A_{ijk}C^{\Gamma_0}_{ijklmn}A_{lmn}, \\ =&\ 2\ell_{(6,3)}(A_{ijk}A_{kij} + A_{ijk}A_{jik}), \\ &+ 2\ell_{(6,4)}A_{ijk}A_{ijk} + \ell_{(6,5)}A_{ikk}A_{imm}. \end{aligned} \tag{301}$$

Lastly, for the continuum type of theory dealing with scalar kinematic processes, expression (296) needs further specialization. Thus, consider that situation where the full minor symmetry is required, that is, $i \leftrightarrow j \leftrightarrow k$. Starting over (296), assume one additional symmetry, $i \leftrightarrow j$, while together with the major symmetry, implies another symmetry, namely, $l \leftrightarrow m$, and we arrive at two independent terms. Then, the tensor C_{ijklmn} can be expressed as

$$\begin{aligned} C^{\Gamma_0}_{ijklmn} &= \ell_{(6,1)}(\delta_{ij}\delta_{kl}\delta_{mn} + \delta_{ij}\delta_{km}\delta_{ln} + \delta_{ik}\delta_{jm}\delta_{ln} \\ &\quad + \delta_{ik}\delta_{jn}\delta_{lm} + \delta_{ik}\delta_{jl}\delta_{mn} + \delta_{ij}\delta_{kn}\delta_{lm} \\ &\quad + \delta_{il}\delta_{jk}\delta_{mn} + \delta_{in}\delta_{jk}\delta_{lm} + \delta_{im}\delta_{jk}\delta_{ln}) \\ &\quad + \ell_{(6,2)}(\delta_{im}\delta_{jn}\delta_{kl} + \delta_{in}\delta_{jm}\delta_{kl} + \delta_{im}\delta_{jl}\delta_{kn} \\ &\quad + \delta_{il}\delta_{jm}\delta_{kn} + \delta_{il}\delta_{jn}\delta_{km} + \delta_{in}\delta_{jl}\delta_{km}). \end{aligned} \tag{302}$$

Contracting the C_{ijklmn} in (302) with A_{ijk}, symmetric under all permutations of i, j, and k, we obtain

$$\begin{aligned} C^{\Gamma_0}_{ijklmn}A_{lmn} &= 3\ell_{(6,1)}(\delta_{ij}A_{kmm} + \delta_{ik}A_{jmm} + \delta_{jk}A_{imm}) \\ &\quad + 6\ell_{(6,2)}A_{ijk}. \end{aligned} \tag{303}$$

Lastly, the norm squared of A_{ijk} becomes

$$\begin{aligned} \|\mathbb{A}\|^2 &:= A_{ijk}C^{\Gamma_0}_{ijklmn}A_{lmn}, \\ &= 9\ell_{(6,1)}A_{kii}A_{kjj} + 6\ell_{(6,2)}A_{ijk}A_{ijk}. \end{aligned} \tag{304}$$

6.4 Eighth-Order Isotropic Tensor

Finally, consider the eighth-order tensor $C_{ijklmnpq}$ where for the general case we have

$$\begin{aligned} \bigotimes_{i=1}^{8}\Gamma_1 &= \Gamma_1 \otimes \Gamma_1 \otimes \Gamma_1 \otimes \Gamma_1 \otimes \Gamma_1 \otimes \Gamma_1 \otimes \Gamma_1 \otimes \Gamma_1, \\ &= (\Gamma_0 \oplus \Gamma_1 \oplus \Gamma_2) \otimes (\Gamma_0 \oplus \Gamma_1 \oplus \Gamma_2) \otimes (\Gamma_0 \oplus \Gamma_1 \oplus \Gamma_2) \otimes (\Gamma_0 \oplus \Gamma_1 \oplus \Gamma_2), \\ &= (15\Gamma_0 \oplus 36\Gamma_1 \oplus 40\Gamma_2 \oplus 29\Gamma_3 \oplus 15\Gamma_4 \oplus 5\Gamma_5 \oplus \Gamma_6) \otimes (\Gamma_0 \oplus 6\Gamma_1 \oplus \Gamma_2), \\ &= 91\Gamma_0 \oplus 232\Gamma_1 \oplus 280\Gamma_2 \oplus 238\Gamma_3 \oplus 154\Gamma_4 \oplus 76\Gamma_5 \oplus 28\Gamma_6 \oplus 7\Gamma_7 \oplus \Gamma_8. \end{aligned} \tag{305}$$

We end up with 91 copies of the trivial irrep Γ_0. Thus, we need 91 coefficients to describe the most general eighth-order isotropic tensor. The blind tensor product combination of four Kronecker deltas provides 105 terms, which are not independent when it comes to considering the SO(3) group. The 105 terms are only independent if we consider rotations in four or higher dimensions. This issue arises because we have four Kronecker delta terms combined where the indices follow an arithmetic modulo three.

To start our discussion on the 91 independent terms, let us first list the 105 terms arising from this blind tensor product combination.

$$
\begin{aligned}
&(\delta_{iq}\delta_{jp}\delta_{kn}\delta_{lm};\ \delta_{ip}\delta_{jm}\delta_{kq}\delta_{ln};\ \delta_{in}\delta_{jk}\delta_{lq}\delta_{mp};\ \delta_{il}\delta_{jp}\delta_{kq}\delta_{mn};\ \delta_{ik}\delta_{jm}\delta_{lq}\delta_{np};\\
&\delta_{iq}\delta_{jp}\delta_{km}\delta_{ln};\ \delta_{ip}\delta_{jm}\delta_{kn}\delta_{lq};\ \delta_{in}\delta_{jk}\delta_{lp}\delta_{mq};\ \delta_{il}\delta_{jp}\delta_{kn}\delta_{mq};\ \delta_{ik}\delta_{jm}\delta_{lp}\delta_{nq};\\
&\delta_{iq}\delta_{jp}\delta_{kl}\delta_{mn};\ \delta_{ip}\delta_{jm}\delta_{kl}\delta_{nq};\ \delta_{in}\delta_{jk}\delta_{lm}\delta_{pq};\ \delta_{il}\delta_{jp}\delta_{km}\delta_{nq};\ \delta_{ik}\delta_{jm}\delta_{ln}\delta_{pq};\\
&\delta_{iq}\delta_{jn}\delta_{kp}\delta_{lm};\ \delta_{ip}\delta_{jl}\delta_{kq}\delta_{mn};\ \delta_{im}\delta_{jq}\delta_{kp}\delta_{ln};\ \delta_{il}\delta_{jn}\delta_{kq}\delta_{mp};\ \delta_{ik}\delta_{jl}\delta_{mq}\delta_{np};\\
&\delta_{iq}\delta_{jn}\delta_{km}\delta_{lp};\ \delta_{ip}\delta_{jl}\delta_{kn}\delta_{mq};\ \delta_{im}\delta_{jq}\delta_{kn}\delta_{lp};\ \delta_{il}\delta_{jn}\delta_{kp}\delta_{mq};\ \delta_{ik}\delta_{jl}\delta_{mp}\delta_{nq};\\
&\delta_{iq}\delta_{jn}\delta_{kl}\delta_{mp};\ \delta_{ip}\delta_{jl}\delta_{km}\delta_{nq};\ \delta_{im}\delta_{jq}\delta_{kl}\delta_{np};\ \delta_{il}\delta_{jn}\delta_{km}\delta_{pq};\ \delta_{ik}\delta_{jl}\delta_{mn}\delta_{pq};\\
&\delta_{iq}\delta_{jm}\delta_{kp}\delta_{ln};\ \delta_{ip}\delta_{jk}\delta_{lq}\delta_{mn};\ \delta_{im}\delta_{jp}\delta_{kq}\delta_{ln};\ \delta_{il}\delta_{jm}\delta_{kq}\delta_{np};\ \delta_{ij}\delta_{kq}\delta_{lp}\delta_{mn};\\
&\delta_{iq}\delta_{jm}\delta_{kn}\delta_{lp};\ \delta_{ip}\delta_{jk}\delta_{ln}\delta_{mq};\ \delta_{im}\delta_{jp}\delta_{kn}\delta_{lq};\ \delta_{il}\delta_{jm}\delta_{kp}\delta_{nq};\ \delta_{ij}\delta_{kq}\delta_{ln}\delta_{mp};\\
&\delta_{iq}\delta_{jm}\delta_{kl}\delta_{np};\ \delta_{ip}\delta_{jk}\delta_{lm}\delta_{nq};\ \delta_{im}\delta_{jp}\delta_{kl}\delta_{nq};\ \delta_{il}\delta_{jm}\delta_{kn}\delta_{pq};\ \delta_{ij}\delta_{kq}\delta_{lm}\delta_{np};\\
&\delta_{iq}\delta_{jl}\delta_{kp}\delta_{mn};\ \delta_{in}\delta_{jq}\delta_{kp}\delta_{lm};\ \delta_{im}\delta_{jn}\delta_{kq}\delta_{lp};\ \delta_{il}\delta_{jk}\delta_{mq}\delta_{np};\ \delta_{ij}\delta_{kp}\delta_{lq}\delta_{mn};\\
&\delta_{iq}\delta_{jl}\delta_{kn}\delta_{mp};\ \delta_{in}\delta_{jq}\delta_{km}\delta_{lp};\ \delta_{im}\delta_{jn}\delta_{kp}\delta_{lq};\ \delta_{il}\delta_{jk}\delta_{mp}\delta_{nq};\ \delta_{ij}\delta_{kp}\delta_{ln}\delta_{mq};\\
&\delta_{iq}\delta_{jl}\delta_{km}\delta_{np};\ \delta_{in}\delta_{jq}\delta_{kl}\delta_{mp};\ \delta_{im}\delta_{jn}\delta_{kl}\delta_{pq};\ \delta_{il}\delta_{jk}\delta_{mn}\delta_{pq};\ \delta_{ij}\delta_{kp}\delta_{lm}\delta_{nq};\\
&\delta_{iq}\delta_{jk}\delta_{lp}\delta_{mn};\ \delta_{in}\delta_{jp}\delta_{kq}\delta_{lm};\ \delta_{im}\delta_{jl}\delta_{kq}\delta_{np};\ \delta_{ik}\delta_{jq}\delta_{lp}\delta_{mn};\ \delta_{ij}\delta_{kn}\delta_{lq}\delta_{mp};\\
&\delta_{iq}\delta_{jk}\delta_{ln}\delta_{mp};\ \delta_{in}\delta_{jp}\delta_{km}\delta_{lq};\ \delta_{im}\delta_{jl}\delta_{kp}\delta_{nq};\ \delta_{ik}\delta_{jq}\delta_{ln}\delta_{mp};\ \delta_{ij}\delta_{kn}\delta_{lp}\delta_{mq};\\
&\delta_{iq}\delta_{jk}\delta_{lm}\delta_{np};\ \delta_{in}\delta_{jp}\delta_{kl}\delta_{mq};\ \delta_{im}\delta_{jl}\delta_{kn}\delta_{pq};\ \delta_{ik}\delta_{jq}\delta_{lm}\delta_{np};\ \delta_{ij}\delta_{kn}\delta_{lm}\delta_{pq};\\
&\delta_{ip}\delta_{jq}\delta_{kn}\delta_{lm};\ \delta_{in}\delta_{jm}\delta_{kq}\delta_{lp};\ \delta_{im}\delta_{jk}\delta_{lq}\delta_{np};\ \delta_{ik}\delta_{jp}\delta_{lq}\delta_{mn};\ \delta_{ij}\delta_{km}\delta_{lq}\delta_{np};\\
&\delta_{ip}\delta_{jq}\delta_{km}\delta_{ln};\ \delta_{in}\delta_{jm}\delta_{kp}\delta_{lq};\ \delta_{im}\delta_{jk}\delta_{lp}\delta_{nq};\ \delta_{ik}\delta_{jp}\delta_{ln}\delta_{mq};\ \delta_{ij}\delta_{km}\delta_{lp}\delta_{nq};\\
&\delta_{ip}\delta_{jq}\delta_{kl}\delta_{mn};\ \delta_{in}\delta_{jm}\delta_{kl}\delta_{pq};\ \delta_{im}\delta_{jk}\delta_{ln}\delta_{pq};\ \delta_{ik}\delta_{jp}\delta_{lm}\delta_{nq};\ \delta_{ij}\delta_{km}\delta_{ln}\delta_{pq};\\
&\delta_{ip}\delta_{jn}\delta_{kq}\delta_{lm};\ \delta_{in}\delta_{jl}\delta_{kq}\delta_{mp};\ \delta_{il}\delta_{jq}\delta_{kp}\delta_{mn};\ \delta_{ik}\delta_{jn}\delta_{lq}\delta_{mp};\ \delta_{ij}\delta_{kl}\delta_{mq}\delta_{np};\\
&\delta_{ip}\delta_{jn}\delta_{km}\delta_{lq};\ \delta_{in}\delta_{jl}\delta_{kp}\delta_{mq};\ \delta_{il}\delta_{jq}\delta_{kn}\delta_{mp};\ \delta_{ik}\delta_{jn}\delta_{lp}\delta_{mq};\ \delta_{ij}\delta_{kl}\delta_{mp}\delta_{nq};\\
&\delta_{ip}\delta_{jn}\delta_{kl}\delta_{mq};\ \delta_{in}\delta_{jl}\delta_{km}\delta_{pq};\ \delta_{il}\delta_{jq}\delta_{km}\delta_{np};\ \delta_{ik}\delta_{jn}\delta_{lm}\delta_{pq};\ \delta_{ij}\delta_{kl}\delta_{mn}\delta_{pq}).
\end{aligned}
\tag{306}
$$

Labeling the terms in (306) from $\langle 0\rangle$ to $\langle 104\rangle$ looping though columns, according to [57], the dependent terms are $\langle 0\rangle$, $\langle 1\rangle$, $\langle 12\rangle$, $\langle 13\rangle$, $\langle 34\rangle$, $\langle 35\rangle$, $\langle 44\rangle$, $\langle 49\rangle$, $\langle 58\rangle$, $\langle 59\rangle$, $\langle 68\rangle$, $\langle 75\rangle$, $\langle 76\rangle$, and $\langle 81\rangle$.

Next, consider the major symmetry needed to impose the symmetry of the inner product $(ijkl) \leftrightarrow (mnpq)$. To enforce this symmetry, we have to combine some of these terms, thereby reducing the number of independent terms from 105 to 65. The minor symmetries $j \leftrightarrow k \leftrightarrow l$, combined with the major symmetry, imply other symmetries, namely, $n \leftrightarrow p \leftrightarrow q$. With these minor symmetries, we have seven remaining terms, thus, we need seven coefficients $\ell_{(8,b)}$, with $1 \le b \le 7$, to describe an eighth-order isotropic tensor with the given symmetries. Altogether, we see how the symmetries reduce the number of independent terms from 105 to 7. Indeed, from (306) while imposing the major symmetry and the mentioned minor symmetries, we have that

$$
\begin{aligned}
C^{\Gamma_0}_{ijklmn} = {} & \ell_{(8,1)}(\delta_{in}\delta_{jk}\delta_{lq}\delta_{mp} + \delta_{in}\delta_{jl}\delta_{kq}\delta_{mp} + \delta_{in}\delta_{jq}\delta_{kl}\delta_{mp} + \delta_{ip}\delta_{jk}\delta_{lq}\delta_{mn} \\
& + \delta_{in}\delta_{jk}\delta_{lp}\delta_{mq} + \delta_{iq}\delta_{jk}\delta_{ln}\delta_{mp} + \delta_{iq}\delta_{jk}\delta_{lp}\delta_{mn} + \delta_{ip}\delta_{jk}\delta_{ln}\delta_{mq} \\
& + \delta_{ip}\delta_{jl}\delta_{kq}\delta_{mn} + \delta_{ip}\delta_{jq}\delta_{kl}\delta_{mn} + \delta_{in}\delta_{jl}\delta_{kp}\delta_{mq} + \delta_{in}\delta_{jp}\delta_{kl}\delta_{mq} \\
& + \delta_{iq}\delta_{jl}\delta_{kn}\delta_{mp} + \delta_{iq}\delta_{jn}\delta_{kl}\delta_{mp} + \delta_{iq}\delta_{jl}\delta_{kp}\delta_{mn} + \delta_{iq}\delta_{jp}\delta_{kl}\delta_{mn} \\
& + \delta_{ip}\delta_{jl}\delta_{kn}\delta_{mq} + \delta_{ip}\delta_{jn}\delta_{kl}\delta_{mq} + \delta_{ik}\delta_{jm}\delta_{lq}\delta_{np} + \delta_{ij}\delta_{km}\delta_{lq}\delta_{np} \\
& + \delta_{il}\delta_{jm}\delta_{kq}\delta_{np} + \delta_{ik}\delta_{jq}\delta_{lm}\delta_{np} + \delta_{ij}\delta_{kq}\delta_{lm}\delta_{np} + \delta_{il}\delta_{jq}\delta_{km}\delta_{np} \\
& + \delta_{ik}\delta_{jm}\delta_{lp}\delta_{nq} + \delta_{ik}\delta_{jm}\delta_{ln}\delta_{pq} + \delta_{ij}\delta_{km}\delta_{lp}\delta_{nq} + \delta_{il}\delta_{jm}\delta_{kp}\delta_{nq} \\
& + \delta_{ik}\delta_{jp}\delta_{lm}\delta_{nq} + \delta_{ij}\delta_{kp}\delta_{lm}\delta_{nq} + \delta_{il}\delta_{jp}\delta_{km}\delta_{nq} + \delta_{ij}\delta_{km}\delta_{ln}\delta_{pq} \\
& + \delta_{il}\delta_{jm}\delta_{kn}\delta_{pq} + \delta_{ik}\delta_{jn}\delta_{lm}\delta_{pq} + \delta_{ij}\delta_{kn}\delta_{lm}\delta_{pq} + \delta_{il}\delta_{jn}\delta_{km}\delta_{pq}) \\
& + \ell_{(8,2)}(\delta_{iq}\delta_{jp}\delta_{kn}\delta_{lm} + \delta_{iq}\delta_{jn}\delta_{kp}\delta_{lm} + \delta_{iq}\delta_{jp}\delta_{km}\delta_{ln} \\
& + \delta_{iq}\delta_{jm}\delta_{kn}\delta_{lp} + \delta_{iq}\delta_{jn}\delta_{km}\delta_{lp} + \delta_{iq}\delta_{jm}\delta_{kp}\delta_{ln} + \delta_{ip}\delta_{jq}\delta_{kn}\delta_{lm} \\
& + \delta_{in}\delta_{jp}\delta_{kq}\delta_{lm} + \delta_{ip}\delta_{jn}\delta_{kq}\delta_{lm} + \delta_{in}\delta_{jq}\delta_{kp}\delta_{lm} + \delta_{ip}\delta_{jq}\delta_{km}\delta_{ln} \\
& + \delta_{ip}\delta_{jm}\delta_{kn}\delta_{lq} + \delta_{ip}\delta_{jn}\delta_{km}\delta_{lq} + \delta_{ip}\delta_{jm}\delta_{kq}\delta_{ln} + \delta_{in}\delta_{jp}\delta_{km}\delta_{lq} \\
& + \delta_{in}\delta_{jm}\delta_{kq}\delta_{lp} + \delta_{in}\delta_{jq}\delta_{km}\delta_{lp} + \delta_{in}\delta_{jm}\delta_{kp}\delta_{lq}) \\
& + \ell_{(8,3)}(\delta_{il}\delta_{jp}\delta_{kq}\delta_{mn} + \delta_{il}\delta_{jq}\delta_{kp}\delta_{mn} + \delta_{ik}\delta_{jp}\delta_{lq}\delta_{mn} + \delta_{ij}\delta_{kq}\delta_{lp}\delta_{mn} \\
& + \delta_{ik}\delta_{jq}\delta_{lp}\delta_{mn} + \delta_{ij}\delta_{kp}\delta_{lq}\delta_{mn} + \delta_{il}\delta_{jn}\delta_{kq}\delta_{mp} + \delta_{il}\delta_{jp}\delta_{kn}\delta_{mq} \\
& + \delta_{il}\delta_{jn}\delta_{kp}\delta_{mq} + \delta_{il}\delta_{jq}\delta_{kn}\delta_{mp} + \delta_{ik}\delta_{jn}\delta_{lq}\delta_{mp} + \delta_{ij}\delta_{kq}\delta_{ln}\delta_{mp} \\
& + \delta_{ik}\delta_{jq}\delta_{ln}\delta_{mp} + \delta_{ij}\delta_{kn}\delta_{lq}\delta_{mp} + \delta_{ik}\delta_{jp}\delta_{ln}\delta_{mq} + \delta_{ij}\delta_{kn}\delta_{lp}\delta_{mq} \\
& + \delta_{ik}\delta_{jn}\delta_{lp}\delta_{mq} + \delta_{ij}\delta_{kp}\delta_{ln}\delta_{mq}) \\
& + \ell_{(8,4)}(\delta_{ip}\delta_{jm}\delta_{kl}\delta_{nq} + \delta_{ip}\delta_{jl}\delta_{km}\delta_{nq} + \delta_{ip}\delta_{jk}\delta_{lm}\delta_{nq} + \delta_{in}\delta_{jm}\delta_{kl}\delta_{pq} \\
& + \delta_{iq}\delta_{jm}\delta_{kl}\delta_{np} + \delta_{in}\delta_{jk}\delta_{lm}\delta_{pq} + \delta_{iq}\delta_{jl}\delta_{km}\delta_{np} + \delta_{iq}\delta_{jk}\delta_{lm}\delta_{np} \\
& + \delta_{in}\delta_{jl}\delta_{km}\delta_{pq}) \\
& + \ell_{(8,5)}(\delta_{ik}\delta_{jl}\delta_{mq}\delta_{np} + \delta_{ij}\delta_{kl}\delta_{mq}\delta_{np} + \delta_{il}\delta_{jk}\delta_{mq}\delta_{np} + \delta_{ik}\delta_{jl}\delta_{mp}\delta_{nq} \\
& + \delta_{ik}\delta_{jl}\delta_{mn}\delta_{pq} + \delta_{ij}\delta_{kl}\delta_{mp}\delta_{nq} + \delta_{ij}\delta_{kl}\delta_{mn}\delta_{pq} + \delta_{il}\delta_{jk}\delta_{mn}\delta_{pq} \\
& + \delta_{il}\delta_{jk}\delta_{mp}\delta_{nq})
\end{aligned}
$$

$$+ \ell_{(8,6)}(\delta_{im}\delta_{jq}\delta_{kl}\delta_{np} + \delta_{im}\delta_{jl}\delta_{kq}\delta_{np} + \delta_{im}\delta_{jk}\delta_{lq}\delta_{np} + \delta_{im}\delta_{jp}\delta_{kl}\delta_{nq}$$
$$+ \delta_{im}\delta_{jn}\delta_{kl}\delta_{pq} + \delta_{im}\delta_{jl}\delta_{kp}\delta_{nq} + \delta_{im}\delta_{jk}\delta_{lp}\delta_{nq} + \delta_{im}\delta_{jl}\delta_{kn}\delta_{pq}$$
$$+ \delta_{im}\delta_{jk}\delta_{ln}\delta_{pq})$$
$$+ \ell_{(8,7)}(\delta_{im}\delta_{jq}\delta_{kp}\delta_{ln} + \delta_{im}\delta_{jp}\delta_{kq}\delta_{ln} + \delta_{im}\delta_{jq}\delta_{kn}\delta_{lp} + \delta_{im}\delta_{jn}\delta_{kp}\delta_{lq}$$
$$+ \delta_{im}\delta_{jp}\delta_{kn}\delta_{lq} + \delta_{im}\delta_{jn}\delta_{kq}\delta_{lp}). \tag{307}$$

So far, we have not considered the dependency among the Kronecker delta products. Thus, the seven terms in (307) may be further reduced. Instead, we select the 14 dependent Kronecker delta products, which are mentioned in the paragraph containing (306), and compute the major and minor symmetries. These symmetries imply that 96 Kronecker delta products, out of 105, are dependent. Thus, we end up with the following nine-independent delta Kronecker products

$$(\delta_{ik}\delta_{jl}\delta_{mn}\delta_{pq};\ \delta_{il}\delta_{jk}\delta_{mp}\delta_{nq};\ \delta_{ij}\delta_{kl}\delta_{mn}\delta_{pq};$$
$$\delta_{il}\delta_{jk}\delta_{mn}\delta_{pq};\ \delta_{ik}\delta_{jl}\delta_{mp}\delta_{nq};\ \delta_{ij}\delta_{kl}\delta_{mp}\delta_{nq}; \tag{308}$$
$$\delta_{il}\delta_{jk}\delta_{mq}\delta_{np};\ \delta_{ik}\delta_{jl}\delta_{mq}\delta_{np};\ \delta_{ij}\delta_{kl}\delta_{mq}\delta_{np}).$$

Note that, all Kronecker delta products in (308) are related by all symmetries here considered.

Contracting $C^{\Gamma_0}_{ijklmnpq}$ with a fourth-order tensor A_{ijkl} that satisfies $A_{ijkl} = A_{ikjl} = A_{ijlk} = A_{ilkj}$, then

$$C^{\Gamma_0}_{ijklmnpq}A_{mnpq} = 3\ell_{(8,1)}(\delta_{ij}A_{lknn} + \delta_{ij}A_{klnn} + \delta_{ik}A_{jlnn} + \delta_{ik}A_{ljnn} + \delta_{il}A_{jknn}$$
$$+ \delta_{il}A_{kjnn} + \delta_{jl}A_{mmik} + \delta_{jl}A_{mmki} + \delta_{jk}A_{mmil} + \delta_{jk}A_{mmli}$$
$$+ \delta_{kl}A_{mmij} + \delta_{kl}A_{mmji})$$
$$+ 6\ell_{(8,2)}(A_{lijk} + A_{jikl} + A_{kijl})$$
$$+ 6\ell_{(8,3)}(\delta_{ij}A_{mmkl} + \delta_{ik}A_{mmjl} + \delta_{il}A_{mmjk})$$
$$+ 3\ell_{(8,4)}(\delta_{jk}A_{limm} + \delta_{jl}A_{kimm} + \delta_{kl}A_{jimm})$$
$$+ 3\ell_{(8,5)}A_{mmpp}(\delta_{ij}\delta_{kl} + \delta_{ik}\delta_{jl} + \delta_{il}\delta_{jk})$$
$$+ 3\ell_{(8,6)}(\delta_{kl}A_{ijnn} + \delta_{jl}A_{ikmm} + \delta_{jk}A_{ilmm})$$
$$+ 6\ell_{(8,7)}A_{ijkl}. \tag{309}$$

Furthermore, the norm squared of A_{ijkl} reads

$$
\begin{aligned}
\|\mathbb{A}\|^2 := & A_{ijkl}C^{\Gamma_0}_{ijklmnpq}A_{mnpq}, \\
= & 9\ell_{(8,1)}(2A_{kkij}A_{ijll} + (A_{kjll} + A_{jkll})A_{iijk}) \\
& + 6\ell_{(8,2)}((A_{lijk} + A_{jikl} + A_{kijl})A_{ijkl}) \\
& + 18\ell_{(8,3)}A_{iikl}A_{jjkl} + 9\ell_{(8,4)}A_{ijkk}A_{jill} \\
& + 9\ell_{(8,5)}A_{iijj}A_{kkll} + 9\ell_{(8,6)}A_{ijkk}A_{ijll} \\
& + 6\ell_{(8,7)}A_{ijkl}A_{ijkl}.
\end{aligned} \tag{310}
$$

Again, for a theory involving incompressible materials, consider that $\mathbb{A}$ satisfies $\mathbf{1} : \mathbb{A} = \mathbf{0}$, then, expressions (309) specializes to

$$
\begin{aligned}
C^{\Gamma_0}_{ijklmnpq}A_{mnpq} = & 3\ell_{(8,1)}(\delta_{ij}A_{lknn} + \delta_{ij}A_{klnn} + \delta_{ik}A_{jlnn} + \delta_{ik}A_{ljnn} + \delta_{il}A_{jknn} \\
& + \delta_{il}A_{kjnn}) \\
& + 6\ell_{(8,2)}(A_{lijk} + A_{jikl} + A_{kijl}) \\
& + 3\ell_{(8,4)}(\delta_{jk}A_{limm} + \delta_{jl}A_{kimm} + \delta_{kl}A_{jimm}) \\
& + 3\ell_{(8,6)}(\delta_{kl}A_{ijnn} + \delta_{jl}A_{ikmm} + \delta_{jk}A_{ilmm}) \\
& + 6\ell_{(8,7)}A_{ijkl}.
\end{aligned} \tag{311}
$$

Letting $S_{ijkl} := C^{\Gamma_0}_{ijklmnpq}A_{mnpq}$, without loss of generality, we impose $\mathbf{1} : \mathbb{S} = \mathbf{0}$ and from (311) arrive at the following condition: $5\ell_{(8,1)} + 2\ell_{(8,2)} + \ell_{(8,4)} + \ell_{(8,6)} = 0$. Moreover, the terms multiplying $\ell_{(8,1)}$, $\ell_{(8,3)}$, and $\ell_{(8,5)}$ are powerless in expression (310). Thus, we opt for $\ell_{(8,2)} = -\frac{1}{2}(5\ell_{(8,1)} + \ell_{(8,4)} + \ell_{(8,6)})$ and expression (311) is specialized to

$$
\begin{aligned}
C^{\Gamma_0}_{ijklmnpq}A_{mnpq} = & 3\ell_{(8,1)}((\delta_{ij}A_{lknn} + \delta_{ij}A_{klnn} + \delta_{ik}A_{jlnn} + \delta_{ik}A_{ljnn} \\
& + \delta_{il}A_{jknn} + \delta_{il}A_{kjnn}) - 5(A_{lijk} + A_{jikl} + A_{kijl})) \\
& + 3\ell_{(8,4)}(\delta_{jk}A_{limm} + \delta_{jl}A_{kimm} + \delta_{kl}A_{jimm} \\
& - (A_{lijk} + A_{jikl} + A_{kijl})) \\
& + 3\ell_{(8,6)}(\delta_{kl}A_{ijnn} + \delta_{jl}A_{ikmm} + \delta_{jk}A_{ilmm} \\
& - (A_{lijk} + A_{jikl} + A_{kijl})) \\
& + 6\ell_{(8,7)}A_{ijkl},
\end{aligned} \tag{312}
$$

while (310) becomes

$$
\begin{aligned}
\|\mathbb{A}\|^2 := {} & A_{ijkl} C^{\Gamma_0}_{ijklmnpq} A_{mnpq}, \\
= {} & -15\ell_{(8,1)} A_{ijkl}(A_{lijk} + A_{jikl} + A_{kijl}) \\
& + 3\ell_{(8,4)}(A_{ijjl}A_{limm} + A_{ijkj}A_{kimm} + A_{ijkk}A_{jimm} \\
& - A_{ijkl}(A_{lijk} + A_{jikl} + A_{kijl})) \\
& + 3\ell_{(8,6)}(A_{ijkk}A_{ijnn} + A_{ijkj}A_{ikmm} + A_{ijjl}A_{ilmm} \\
& - A_{ijkl}(A_{lijk} + A_{jikl} + A_{kijl})) \\
& + 6\ell_{(8,7)} A_{ijkl}A_{ijkl}. \qquad (313)
\end{aligned}
$$

Synopsis, chapter 5

(i) Partwise free-energy imbalance for isothermal processes for a spatial part $\mathcal{P}_\tau$*:*

$$\overline{\int_{\mathcal{P}_\tau} \varrho\psi \, \mathrm{d}v_\tau}^{\boldsymbol{\cdot}} - \mathcal{W}_{\text{ext}}(\mathcal{P}_\tau) =: -\int_{\mathcal{P}_\tau} \delta \, \mathrm{d}v_\tau \le 0$$

(ii) Partwise free-energy imbalance for isothermal processes for a control volume $\overline{\mathcal{P}}$*:*

$$\overline{\int_{\overline{\mathcal{P}}} \varrho\psi \, \mathrm{d}v}^{\boldsymbol{\cdot}} + \int_{\partial\overline{\mathcal{P}}} \varrho\psi \, \boldsymbol{v} \cdot \boldsymbol{n} \, \mathrm{d}v - \mathcal{W}_{\text{ext}}(\overline{\mathcal{P}}) =: -\int_{\overline{\mathcal{P}}} \delta \, \mathrm{d}v \le 0$$

(iii) Pointwise free-energy imbalance for isothermal processes:

$$\begin{aligned}
0 \ge \varrho\dot{\psi} &- (\mathbf{T} + \sum_{\alpha=1}^{n} (\operatorname{grad}\varphi^\alpha \otimes \boldsymbol{\xi}^\alpha + 2(\operatorname{grad}^2\varphi^\alpha)\boldsymbol{\Xi}^\alpha + 3\operatorname{grad}^3\varphi^\alpha : \mathbb{X}^\alpha)) : \operatorname{grad}\boldsymbol{v} \\
&- (\mathbb{T} + \sum_{\alpha=1}^{n} (\operatorname{grad}\varphi^\alpha \otimes \boldsymbol{\Xi}^\alpha + 3\operatorname{grad}^2\varphi^\alpha \cdot \mathbb{X}^\alpha)) \vdots \operatorname{grad}^2\boldsymbol{v} \\
&- (\mathbf{T} + \sum_{\alpha=1}^{n} \operatorname{grad}\varphi^\alpha \otimes \mathbb{X}^\alpha) :: \operatorname{grad}^3\boldsymbol{v} \\
&+ \sum_{\alpha=1}^{n} \Big\{ \pi^\alpha \dot{\varphi}^\alpha - \boldsymbol{\xi}^\alpha \cdot (\operatorname{grad}\varphi^\alpha)^{\boldsymbol{\cdot}} - \boldsymbol{\Xi}^\alpha : (\operatorname{grad}^2\varphi^\alpha)^{\boldsymbol{\cdot}} - \mathbb{X}^\alpha \vdots (\operatorname{grad}^3\varphi^\alpha)^{\boldsymbol{\cdot}} \Big\}
\end{aligned}$$

Representation theorems

(i) Isotropic fourth-order $C^{\Gamma_0}_{ijkl}$*:* Major $(ij) \leftrightarrow (kl)$ and minor $i \leftrightarrow j$ symmetries render two Clebsch–Gordan coefficients $\ell_{(4,b)}$ with $1 \le b \le 2$.

(ii) Isotropic fourth-order $C^{\Gamma_0}_{ijkl}$ *tensor acting on a second-order tensor* A_{kl}*:*

$$C^{\Gamma_0}_{ijkl} A_{kl} = \ell_{(4,1)} \delta_{ij} A_{kk} + 2\ell_{(4,2)} A_{ij}$$

(iii) Norm squared of a second-order tensor A_{kl}*:*

$$\begin{aligned}
\|\mathbf{A}\|^2 &:= A_{ij} C^{\Gamma_0}_{ijkl} A_{kl} \\
&= \ell_{(4,1)} A_{ii} A_{kk} + 2\ell_{(4,2)} A_{ij} A_{ij}
\end{aligned}$$

(iv) Isotropic fourth-order $C^{\Gamma_0}_{ijkl}$ tensor acting on a traceless second-order tensor A_{kl}:

$$C^{\Gamma_0}_{ijkl} A_{kl} = 2\ell_{(4,2)} A_{ij}$$

(v) Norm squared of a traceless second-order tensor A_{kl}:

$$\begin{aligned}\|\mathbf{A}\|^2 &:= A_{ij} C^{\Gamma_0}_{ijkl} A_{kl} \\ &= 2\ell_{(4,2)} A_{ij} A_{ij}\end{aligned}$$

(i) Isotropic sixth-order $C^{\Gamma_0}_{ijklmn}$: Major $(ijk) \leftrightarrow (lmn)$ and minor $i \leftrightarrow j \leftrightarrow k$ symmetries render two Clebsch–Gordan coefficients $\ell_{(6,b)}$ with $1 \leq b \leq 2$.

(ii) Isotropic sixth-order $C^{\Gamma_0}_{ijklmn}$ tensor acting on a third-order tensor A_{lmn}:

$$\begin{aligned}C^{\Gamma_0}_{ijklmn} A_{lmn} &= 3\ell_{(6,1)}(\delta_{ij} A_{kmm} + \delta_{ik} A_{jmm} + \delta_{jk} A_{imm}) \\ &\quad + 6\ell_{(6,2)} A_{ijk}\end{aligned}$$

(iii) Norm squared of a third-order tensor A_{lmn}:

$$\begin{aligned}\|\mathbb{A}\|^2 &:= A_{ijk} C^{\Gamma_0}_{ijklmn} A_{lmn} \\ &= 9\ell_{(6,1)} A_{kii} A_{kjj} + 6\ell_{(6,2)} A_{ijk} A_{ijk}\end{aligned}$$

(i) Isotropic sixth-order $C^{\Gamma_0}_{ijklmn}$: Major $(ijk) \leftrightarrow (lmn)$ and minor $j \leftrightarrow k$ symmetries render five Clebsch–Gordan coefficients $\ell_{(6,b)}$ with $1 \leq b \leq 5$.

(ii) Isotropic sixth-order $C^{\Gamma_0}_{ijklm}$ tensor acting on a traceless third-order tensor A_{lmn}:

$$\begin{aligned}C^{\Gamma_0}_{ijklmn} A_{lmn} &= 2\ell_{(6,3)}(A_{kij} + A_{jik} - \tfrac{1}{2}(\delta_{ij} A_{kmm} + \delta_{ik} A_{jmm})) \\ &\quad + 2\ell_{(6,4)} A_{ijk} + \ell_{(6,5)}(\delta_{jk} A_{imm} - \tfrac{1}{2}(\delta_{ij} A_{kmm} + \delta_{ik} A_{jmm}))\end{aligned}$$

(iii) Norm squared of a traceless third-order tensor A_{lmn}:

$$\begin{aligned}\|\mathbb{A}\|^2 &:= A_{ijk} C^{\Gamma_0}_{ijklmn} A_{lmn} \\ &= 2\ell_{(6,3)}(A_{ijk} A_{kij} + A_{ijk} A_{jik}) \\ &\quad + 2\ell_{(6,4)} A_{ijk} A_{ijk} + \ell_{(6,5)} A_{ikk} A_{imm}\end{aligned}$$

(i) Isotropic eighth-order $C^{\Gamma_0}_{ijklmnpq}$*:* Major $(ijkl) \leftrightarrow (mnpq)$ and minor $j \leftrightarrow k \leftrightarrow l$ symmetries render seven Clebsch–Gordan coefficients $\ell_{(8,b)}$ with $1 \leq b \leq 7$.

(ii) Isotropic eighth-order $C^{\Gamma_0}_{ijklmnpq}$ *tensor acting on a traceless fourth-order tensor* A_{mnpq}*:*

$$
\begin{aligned}
C^{\Gamma_0}_{ijklmnpq} A_{mnpq} = {} & 3\ell_{(8,1)}((\delta_{ij} A_{lknn} + \delta_{ij} A_{klnn} + \delta_{ik} A_{jlnn} + \delta_{ik} A_{ljnn} \\
& + \delta_{il} A_{jknn} + \delta_{il} A_{kjnn}) - 5(A_{lijk} + A_{jikl} + A_{kijl})) \\
& + 3\ell_{(8,4)}(\delta_{jk} A_{limm} + \delta_{jl} A_{kimm} + \delta_{kl} A_{jimm} \\
& - (A_{lijk} + A_{jikl} + A_{kijl})) \\
& + 3\ell_{(8,6)}(\delta_{kl} A_{ijnn} + \delta_{jl} A_{ikmm} + \delta_{jk} A_{ilmm} \\
& - (A_{lijk} + A_{jikl} + A_{kijl})) \\
& + 6\ell_{(8,7)} A_{ijkl}
\end{aligned}
$$

(iii) Norm squared of a traceless fourth-order tensor A_{mnpq}*:*

$$
\begin{aligned}
\|\mathbb{A}\|^2 := {} & A_{ijkl} C^{\Gamma_0}_{ijklmnpq} A_{mnpq} \\
= {} & -15\ell_{(8,1)} A_{ijkl}(A_{lijk} + A_{jikl} + A_{kijl}) \\
& + 3\ell_{(8,4)}(A_{ijjl} A_{limm} + A_{ijkj} A_{kimm} + A_{ijkk} A_{jimm} \\
& - A_{ijkl}(A_{lijk} + A_{jikl} + A_{kijl})) \\
& + 3\ell_{(8,6)}(A_{ijkk} A_{ijnn} + A_{ijkj} A_{ikmm} + A_{ijjl} A_{ilmm} \\
& - A_{ijkl}(A_{lijk} + A_{jikl} + A_{kijl})) \\
& + 6\ell_{(8,7)} A_{ijkl} A_{ijkl}
\end{aligned}
$$

Chapter 6
Coupling

Until this chapter, we have mainly studied balances that are agnostic to material idealizations. Thus, our balances may be used for different solid and fluid materials. In this chapter, based on the Coleman–Noll procedure, we augment the system of equations with appropriate constitutive relations. Moreover, we aim to determine the constitutive equations for certain particular types of couplings between motion and multiple transition layers. That is, these transition layers may represent nonconserved or conserved quantities. To this end, we rely on the thermodynamics of continua and the Larché & Cahn derivatives.

1 Nonconserved Coupling with an Incompressible Fluid

We hereafter restrict attention to purely mechanical processes governed by the isothermal version (248) of the pointwise free-energy imbalance. Guided by the presence of the power conjugate pairings $\mathbf{T} : \mathrm{grad}\, \boldsymbol{v}$, $\mathbb{T} \vdots \mathrm{grad}^2 \boldsymbol{v}$, $\mathbf{T} :: \mathrm{grad}^3 \boldsymbol{v}$, $\pi^\alpha \dot{\varphi}^\alpha$, $\boldsymbol{\xi}^\alpha \cdot \mathrm{grad}\, \dot{\varphi}^\alpha$, $\Xi^\alpha : \mathrm{grad}^2 \dot{\varphi}^\alpha$, and $\mathbb{X}^\alpha \vdots \mathrm{grad}^3 \dot{\varphi}^\alpha$ in that inequality, we consider a class of constitutive equations that delivers the free-energy density ψ, stress $\mathbf{T}$, hyperstress $\mathbb{T}$, supra hyperstress $\mathbf{T}$, αth internal microforce π^α, αth microstress $\boldsymbol{\xi}^\alpha$, αth hypermicrostress Ξ^α, and αth supra hypermicrostress $\mathbb{X}^\alpha$ at each point $\boldsymbol{x}$ in $\mathcal{B}$ and each instant t of time, in terms of the values of the velocity $\boldsymbol{v}$, its first, second, and third gradients $\mathrm{grad}\, \boldsymbol{v}$, $\mathrm{grad}^2 \boldsymbol{v}$, and $\mathrm{grad}^3 \boldsymbol{v}$, and its time rate $\dot{\boldsymbol{v}}$ as well as in terms of the values of the αth phase field φ^α, its first, second, and third gradients $\mathrm{grad}\, \varphi^\alpha$, $\mathrm{grad}^2 \varphi^\alpha$, and $\mathrm{grad}^3 \varphi^\alpha$, and its time rate $\dot{\varphi}^\alpha$ at that point and time.

The incompressibility assumption, that is, $\varrho = \text{constant}$ and $\psi = \text{constant}$ for φ^α fixed for all $1 \leq \alpha \leq n$, leads us to the following condition

$$\mathrm{div}\, \boldsymbol{v} = \mathrm{tr}\, (\mathrm{grad}\, \boldsymbol{v}) = 0. \tag{314}$$

L. Espath, *Mechanics and Geometry of Enriched Continua*,
https://doi.org/10.1007/978-3-031-28934-7_6

Given the constraint (314), we can establish that there is no spherical contribution to the power expenditure carried out by the stress, hyperstress, and supra hyperstress. Thus, considering the incompressibility condition, we have that

$$\begin{cases} \mathbf{T} : \operatorname{grad} \boldsymbol{v} = (\mathbf{T}_0 - \mathbf{1}g) : \operatorname{grad} \boldsymbol{v} = \mathbf{T}_0 : \operatorname{grad} \boldsymbol{v}, \\ \mathbb{T} \vdots \operatorname{grad}^2 \boldsymbol{v} = (\mathbb{T}_0 - \mathbf{1} \otimes \boldsymbol{g}) \vdots \operatorname{grad}^2 \boldsymbol{v} = \mathbb{T}_0 \vdots \operatorname{grad}^2 \boldsymbol{v}, \\ \mathbf{T} :: \operatorname{grad}^3 \boldsymbol{v} = (\mathbf{T}_0 - \mathbf{1} \otimes \mathbf{G}) :: \operatorname{grad}^3 \boldsymbol{v} = \mathbf{T}_0 :: \operatorname{grad}^3 \boldsymbol{v}, \end{cases} \tag{315}$$

where the indeterminate fields g, $\boldsymbol{g}$, and $\mathbf{G}$ are, respectively, the pressure, hyperpressure, and supra hyperpressure. Thus, $\mathbf{1}g$, $\mathbf{1} \otimes \mathbf{g}$, and $\mathbf{1} \otimes \mathbf{G}$ constitute, respectively, the spherical components of the stress, hyperstress, and supra hyperstress. These pressure-like fields are not thermodynamical pressures since they are powerless and do not provoke changes in volume. Also, note that $\mathbf{1} : \mathbf{T}_0 = 0$, $\mathbf{1} : \mathbb{T}_0 = \mathbf{0}$, and $\mathbf{1} : \mathbf{T}_0 = \mathbf{0}$. Thus, $\mathbf{T}_0$, $\mathbb{T}_0$, and $\mathbf{T}_0$ constitute, respectively, the deviatoric components of the stress, hyperstress, and supra hyperstress.

For the free-energy density ψ given by a constitutive response function $\hat{\psi}$:

$$\psi = \hat{\psi}(\{\varphi^\alpha, \operatorname{grad} \varphi^\alpha, \operatorname{grad}^2 \varphi^\alpha, \operatorname{grad}^3 \varphi^\alpha\}_{\alpha=1}^n), \tag{316}$$

with the incompressibility decomposition of the stress, hyperstress, and supra hyperstress (315), the pointwise free-energy imbalance (253) can be expressed as

$$\begin{aligned} 0 \geq &- (\mathbf{T}_0 + \sum_{\alpha=1}^n (\operatorname{grad} \varphi^\alpha \otimes \boldsymbol{\xi}^\alpha + 2(\operatorname{grad}^2 \varphi) \Xi^\alpha + 3 \operatorname{grad}^3 \varphi : \mathbb{X}^\alpha)) : \operatorname{grad} \boldsymbol{v} \\ &- (\mathbb{T}_0 + \sum_{\alpha=1}^n (\operatorname{grad} \varphi \otimes \Xi^\alpha + 3 \operatorname{grad}^2 \varphi \cdot \mathbb{X}^\alpha)) \vdots \operatorname{grad}^2 \boldsymbol{v} \\ &- (\mathbf{T}_0 + \sum_{\alpha=1}^n \operatorname{grad} \varphi \otimes \mathbb{X}^\alpha) :: \operatorname{grad}^3 \boldsymbol{v} \\ &+ \sum_{\alpha=1}^n \Big\{ (\varrho \partial_{\varphi^\alpha} \hat{\psi} + \pi^\alpha) \dot{\varphi}^\alpha + (\varrho \partial_{\operatorname{grad} \varphi^\alpha} \hat{\psi} - \boldsymbol{\xi}^\alpha) \cdot (\operatorname{grad} \varphi^\alpha)^{\cdot} \\ &+ (\varrho \partial_{\operatorname{grad}^2 \varphi^\alpha} \hat{\psi} - \Xi^\alpha) : (\operatorname{grad}^2 \varphi^\alpha)^{\cdot} + (\varrho \partial_{\operatorname{grad}^3 \varphi^\alpha} \hat{\psi} - \mathbb{X}^\alpha) \vdots (\operatorname{grad}^3 \varphi^\alpha)^{\cdot} \Big\}. \end{aligned} \tag{317}$$

For simplicity, we assume all response functions are isotropic functions. Arguments introduced by Coleman & Noll [59] can then be adapted to show that for the dissipation inequality (317) to be satisfied in all processes it is necessary and sufficient to require that:

- The αth microstress $\boldsymbol{\xi}^\alpha$, αth hypermicrostress Ξ^α, and αth supra hypermicrostress $\mathbb{X}^\alpha$ are given by constitutive response functions $\hat{\boldsymbol{\xi}}^\alpha$, $\hat{\Xi}^\alpha$, and $\hat{\mathbb{X}}^\alpha$ that derive from the response function $\hat{\psi}$:

$$\begin{cases} \boldsymbol{\xi}^\alpha := \hat{\boldsymbol{\xi}}^\alpha(\{\varphi^\alpha, \operatorname{grad}\varphi^\alpha, \operatorname{grad}^2\varphi^\alpha, \operatorname{grad}^3\varphi^\alpha\}_{\alpha=1}^n), \\ \quad = \varrho\partial_{\operatorname{grad}\varphi^\alpha}\hat{\psi}, \\ \Xi^\alpha := \hat{\Xi}^\alpha(\{\varphi^\alpha, \operatorname{grad}\varphi^\alpha, \operatorname{grad}^2\varphi^\alpha, \operatorname{grad}^3\varphi^\alpha\}_{\alpha=1}^n), \\ \quad = \varrho\partial_{\operatorname{grad}^2\varphi^\alpha}\hat{\psi}, \\ \mathbb{X}^\alpha := \hat{\mathbb{X}}^\alpha(\{\varphi^\alpha, \operatorname{grad}\varphi^\alpha, \operatorname{grad}^2\varphi^\alpha, \operatorname{grad}^3\varphi^\alpha\}_{\alpha=1}^n), \\ \quad = \varrho\partial_{\operatorname{grad}^3\varphi^\alpha}\hat{\psi}. \end{cases} \tag{318}$$

- The αth internal microforce π^α is given by a constitutive response function $\hat{\pi}^\alpha$ that splits additively into a contribution derived from the response function $\hat{\psi}$ and a dissipative contribution that, in contrast to $\hat{\psi}$, $\hat{\boldsymbol{\xi}}^\alpha$, $\hat{\Xi}^\alpha$, and $\hat{\mathbb{X}}^\alpha$ depends on $\dot{\varphi}^\alpha$ and must be consistent with a residual dissipation inequality:

$$\begin{cases} \pi^\alpha := \hat{\pi}^\alpha(\{\dot{\varphi}^\alpha, \varphi^\alpha, \operatorname{grad}\varphi^\alpha, \operatorname{grad}^2\varphi^\alpha, \operatorname{grad}^3\varphi^\alpha\}_{\alpha=1}^n), \\ \quad = -\varrho\partial_{\varphi^\alpha}\hat{\psi} + \pi^\alpha_{\text{dis}}(\{\dot{\varphi}^\alpha, \varphi^\alpha, \operatorname{grad}\varphi^\alpha, \operatorname{grad}^2\varphi^\alpha, \operatorname{grad}^3\varphi^\alpha\}_{\alpha=1}^n), \\ \pi^\alpha_{\text{dis}}(\{\dot{\varphi}^\alpha, \varphi^\alpha, \operatorname{grad}\varphi^\alpha, \operatorname{grad}^2\varphi^\alpha, \operatorname{grad}^3\varphi^\alpha\}_{\alpha=1}^n)\dot{\varphi}^\alpha \leq 0. \end{cases} \tag{319}$$

- The stress $\mathbf{T}_0$, the hyperstress $\mathbb{T}_0$, and the supra hyperstress $\mathbf{T}_0$, are, respectively, given by

$$\mathbf{S} = \mathbf{T}_0 + \sum_{\alpha=1}^n (\operatorname{grad}\varphi^\alpha \otimes \boldsymbol{\xi}^\alpha + 2(\operatorname{grad}^2\varphi^\alpha)\Xi^\alpha + 3\operatorname{grad}^3\varphi^\alpha : \mathbb{X}^\alpha), \tag{320}$$

$$\mathbb{S} = \mathbb{T}_0 + \sum_{\alpha=1}^n (\operatorname{grad}\varphi^\alpha \otimes \Xi^\alpha + 3\operatorname{grad}^2\varphi^\alpha \cdot \mathbb{X}^\alpha), \tag{321}$$

and

$$\mathbf{S} = \mathbf{T}_0 + \sum_{\alpha=1}^n \operatorname{grad}\varphi^\alpha \otimes \mathbb{X}^\alpha, \tag{322}$$

with $\mathbf{S} := \hat{\mathbf{S}}(\operatorname{grad}\boldsymbol{v})$, $\mathbb{S} := \hat{\mathbb{S}}(\operatorname{grad}^2\boldsymbol{v})$, and $\mathbf{S} := \hat{\mathbf{S}}(\operatorname{grad}^3\boldsymbol{v})$ representing the thermodynamic stress, hyperstress, and supra hyperstress as they are, respectively, conjugate to $\mathbf{B} := \operatorname{sym}(\operatorname{grad}\boldsymbol{v})$, $\mathbb{B} := \operatorname{grad}^2\boldsymbol{v}$, and $\mathbf{B} := \operatorname{grad}^3\boldsymbol{v}$. Given our choice for the constitutive response function (316), the thermodynamic stress, hyperstress, and supra hyperstress fields are not defined through the free-energy, but still defined by the free-energy imbalance (316).

The simplest choices for the deviatoric components of the stress, hyperstress, and supra hyperstress are given, respectively, by the representation theorem's expressions (292), (300), and (312)

$$(\mathbf{S})_{ij} = 2\ell_{(4,1)}(\mathbf{B})_{ij}, \tag{323}$$

$$\begin{aligned}(\mathbb{S})_{ijk} = {} & 2\ell_{(6,1)}((\mathbb{B})_{kij} + (\mathbb{B})_{jik} - \tfrac{1}{2}(\delta_{ij}(\mathbb{B})_{kmm} + \delta_{ik}(\mathbb{B})_{jmm})) \\ & + 2\ell_{(6,2)}(\mathbb{B})_{ijk} + \ell_{(6,3)}(\delta_{jk}(\mathbb{B})_{imm} - \tfrac{1}{2}(\delta_{ij}(\mathbb{B})_{kmm} + \delta_{ik}(\mathbb{B})_{jmm})),\end{aligned} \tag{324}$$

and

$$\begin{aligned}(\boldsymbol{\mathbb{S}})_{ijkl} = {} & 3\ell_{(8,1)}((\delta_{ij}(\boldsymbol{\mathbb{B}})_{lknn} + \delta_{ij}(\boldsymbol{\mathbb{B}})_{klnn} + \delta_{ik}(\boldsymbol{\mathbb{B}})_{jlnn} + \delta_{ik}(\boldsymbol{\mathbb{B}})_{ljnn} \\ & + \delta_{il}(\boldsymbol{\mathbb{B}})_{jknn} + \delta_{il}(\boldsymbol{\mathbb{B}})_{kjnn}) - 5((\boldsymbol{\mathbb{B}})_{lijk} + (\boldsymbol{\mathbb{B}})_{jikl} + (\boldsymbol{\mathbb{B}})_{kijl})) \\ & + 3\ell_{(8,2)}(\delta_{jk}(\boldsymbol{\mathbb{B}})_{limm} + \delta_{jl}(\boldsymbol{\mathbb{B}})_{kimm} + \delta_{kl}(\boldsymbol{\mathbb{B}})_{jimm} \\ & - ((\boldsymbol{\mathbb{B}})_{lijk} + (\boldsymbol{\mathbb{B}})_{jikl} + (\boldsymbol{\mathbb{B}})_{kijl})) \\ & + 3\ell_{(8,3)}(\delta_{kl}(\boldsymbol{\mathbb{B}})_{ijnn} + \delta_{jl}(\boldsymbol{\mathbb{B}})_{ikmm} + \delta_{jk}(\boldsymbol{\mathbb{B}})_{ilmm} \\ & - ((\boldsymbol{\mathbb{B}})_{lijk} + (\boldsymbol{\mathbb{B}})_{jikl} + (\boldsymbol{\mathbb{B}})_{kijl})) \\ & + 6\ell_{(8,4)}(\boldsymbol{\mathbb{B}})_{ijkl}.\end{aligned} \tag{325}$$

Also, note that we have relabeled the Clebsch–Gordan coefficients in (323), (324), and (325). The only requirement in these constitutive relations is a set of Clebsch–Gordan coefficients which guarantee that expressions (293), (301), and (313) are strictly positive for nonnull tensors $\mathbf{B}$, $\mathbb{B}$, and $\boldsymbol{\mathbb{B}}$, respectively. In view of the constitutive restrictions (316)–(319), the response function for the free-energy density serves as a thermodynamic potential for the αth microstress, the αth hypermicrostress, and the αth supra hypermicrostress, and the equilibrium contribution to the αth internal microforce. A complete description of the response of a material belonging to the class in question thus consists of providing scalar-valued response functions $\hat{\psi}$ and π^α_{dis}. While $\hat{\psi}$ depends only on φ^α, $\operatorname{grad}\varphi^\alpha$, $\operatorname{grad}^2\varphi^\alpha$, and $\operatorname{grad}^3\varphi^\alpha$, π^α_{dis} also depends on $\dot{\varphi}$. Moreover, π^α_{dis} must satisfy the residual dissipation inequality $(319)_2$ for all choices of φ^α, $\operatorname{grad}\varphi^\alpha$, $\operatorname{grad}^2\varphi^\alpha$, $\operatorname{grad}^3\varphi^\alpha$, and $\dot{\varphi}^\alpha$. Consider the simplest response free-energy density ψ (316), composed by quadratic forms

$$\psi = f(\{\varphi^\alpha\}^n_{\alpha=1}) + \tfrac{1}{2}\sum_{\alpha=1}^{n}(\|\boldsymbol{y}^\alpha\|^2 + \|\boldsymbol{\Upsilon}^\alpha\|^2 + \|\mathbb{Y}^\alpha\|^2), \tag{326}$$

where f is a potential function, $\boldsymbol{y}^\alpha := \operatorname{grad}\varphi^\alpha$, $\boldsymbol{\Upsilon}^\alpha := \operatorname{grad}^2\varphi^\alpha$, $\mathbb{Y}^\alpha := \operatorname{grad}^3\varphi^\alpha$, and the quadratic forms are endowed with the metrics $C^{\Gamma_0}_{ij}$, $C^{\Gamma_0}_{ijkl}$, and $C^{\Gamma_0}_{ijklmn}$, respectively. Thus, from the constitutive responses (318), we have

$$(\boldsymbol{\xi}^\alpha)_i = \varrho\ell^\alpha_{(2,1)}(\boldsymbol{y}^\alpha)_i, \tag{327}$$

$$(\boldsymbol{\Xi}^\alpha)_{ij} = \varrho\ell^\alpha_{(4,1)}\delta_{ij}(\boldsymbol{\Upsilon}^\alpha)_{kk} + 2\varrho\ell^\alpha_{(4,2)}(\boldsymbol{\Upsilon}^\alpha)_{ij}, \tag{328}$$

and

$$\begin{aligned}(\mathbb{X}^\alpha)_{ijk} = {} & 3\varrho\ell^\alpha_{(6,1)}(\delta_{ij}(\mathbb{Y}^\alpha)_{kmm} + \delta_{ik}(\mathbb{Y}^\alpha)_{jmm} + \delta_{jk}(\mathbb{Y}^\alpha)_{imm}) \\ & + 6\varrho\ell^\alpha_{(6,2)}(\mathbb{Y}^\alpha)_{ijk},\end{aligned} \tag{329}$$

where expressions (290) and (303) were used to arrive at (328) and (329), respectively.

Using (318) and (319) in the field equations (121), we obtain the αth nonconserved field equation

$$-\pi^\alpha_{\text{dis}} = \operatorname{div}\Big(\varrho\partial_{\operatorname{grad}\varphi^\alpha}\hat{\psi} - \operatorname{div}\Big(\varrho\partial_{\operatorname{grad}^2\varphi^\alpha}\hat{\psi} - \operatorname{div}(\varrho\partial_{\operatorname{grad}^3\varphi^\alpha}\hat{\psi})\Big)\Big) - \varrho\partial_{\varphi^\alpha}\hat{\psi} + \gamma^\alpha, \tag{330}$$

for the phase field, where the simplest choice of π^α_{dis} is $-\beta^\alpha\dot{\varphi}^\alpha$ with $\beta^\alpha \geq 0$ for all $1 \leq \alpha \leq n$. Also, note that $\partial_{\varphi^\alpha}\hat{\psi} = \partial_{\varphi^\alpha} f$. We refer to (330) as the 'αth nonconserved third-grade phase-field equation'.

2 Conserved Coupling with an Incompressible Fluid

We here extend our theory to the case where the phase fields represent the concentration of conserved species while continuing to restrict attention to isothermal processes. The αth concentration is

$$\varphi^\alpha := \varrho^\alpha/\varrho, \tag{331}$$

and the other relevant quantities are the αth chemical potential μ^α, the αth species flux $\jmath^\alpha$, and the αth rate of species production s^α. Moreover, the rate of species production splits additively as

$$s^\alpha := s^\alpha_{\text{ext}} + s^\alpha_{\text{int}}, \tag{332}$$

where s^α_{ext} are the external rates of species production and s^α_{int} the internal (or reactive) rates of species production. We also consider the following constraints

$$\sum_{\alpha=1}^{n}\varphi^\alpha = 1, \qquad \sum_{\alpha=1}^{n} s^\alpha_{\text{ext}} = \sum_{\alpha=1}^{n} s^\alpha_{\text{int}} = 0, \qquad \text{and} \qquad \sum_{\alpha=1}^{n}\jmath^\alpha = \mathbf{0}. \tag{333}$$

Following Gurtin's derivation of the Cahn–Hilliard equation [31, §3], we therefore supplement the field equations (121) by a partwise species balance

$$\overline{\int\limits_{\mathcal{P}_\tau} \varrho\varphi^\alpha \, \mathrm{d}v_\tau}^{\boldsymbol{\cdot}} = \int\limits_{\mathcal{P}_\tau} s^\alpha \, \mathrm{d}v_\tau - \int\limits_{\partial\mathcal{P}_\tau} \boldsymbol{\jmath}^\alpha \cdot \boldsymbol{n} \, \mathrm{d}a_\tau. \tag{334}$$

After localizing it, we obtain the pointwise version of the species balance

$$\varrho\dot{\varphi}^\alpha = s^\alpha - \operatorname{div} \boldsymbol{\jmath}^\alpha. \tag{335}$$

Moreover, we augment the partwise free-energy imbalance (249) to account for the rate at which energy is transferred to $\mathcal{P}_\tau$ due to species transport, yielding

$$\begin{aligned} &\overline{\int\limits_{\mathcal{P}_\tau} \varrho\psi \, \mathrm{d}v_\tau}^{\boldsymbol{\cdot}} - \mathcal{W}_{\mathrm{ext}}(\mathcal{P}_\tau) \\ &\qquad + \sum_{\alpha=1}^{n} \Big\{ - \int\limits_{\mathcal{P}_\tau} \mu^\alpha s^\alpha_{\mathrm{ext}} \, \mathrm{d}v_\tau + \int\limits_{\partial\mathcal{P}_\tau} \mu^\alpha \boldsymbol{\jmath}^\alpha \cdot \boldsymbol{n} \, \mathrm{d}a_\tau \Big\} \\ &\qquad\qquad =: - \int\limits_{\mathcal{P}_\tau} \delta \, \mathrm{d}v_\tau \le 0. \end{aligned} \tag{336}$$

Localizing (336) while using the field equations (121) and the pointwise species balance (335) to eliminate, respectively, the αth external microforce γ^α and the αth external rate of species production s^α_{ext}, we arrive at the pointwise free-energy imbalance

$$\begin{aligned} &\varrho\dot{\psi} - \mathbf{T} : \operatorname{grad} \boldsymbol{v} - \mathbb{T} \,\vdots\, \operatorname{grad}^2 \boldsymbol{v} - \mathbf{T} :: \operatorname{grad}^3 \boldsymbol{v} \\ &\quad + \sum_{\alpha=1}^{n} \Big\{ (\pi^\alpha - \varrho\mu^\alpha)\dot{\varphi}^\alpha - \boldsymbol{\xi}^\alpha \cdot \operatorname{grad} \dot{\varphi}^\alpha - \boldsymbol{\Xi}^\alpha : \operatorname{grad}^2 \dot{\varphi}^\alpha - \mathbb{X}^\alpha \,\vdots\, \operatorname{grad} \dot{\varphi}^\alpha \\ &\qquad + \mu^\alpha s^\alpha_{\mathrm{int}} + \boldsymbol{\jmath}^\alpha \cdot \operatorname{grad} \mu^\alpha \Big\} =: -\delta \le 0, \end{aligned} \tag{337}$$

and with identities (250)–(252) to commutate the material time derivative with the spatial gradients, the pointwise free-energy imbalance (337) assumes the form

$$
\begin{aligned}
0 \geq\ & \varrho\dot{\psi} \\
& - (\mathbf{T} + \sum_{\alpha=1}^{n} (\operatorname{grad}\varphi^\alpha \otimes \boldsymbol{\xi}^\alpha + 2(\operatorname{grad}^2\varphi^\alpha)\boldsymbol{\Xi}^\alpha + 3\operatorname{grad}^3\varphi^\alpha : \mathbb{X}^\alpha)) : \operatorname{grad}\boldsymbol{v} \\
& - (\mathbb{T} + \sum_{\alpha=1}^{n} (\operatorname{grad}\varphi^\alpha \otimes \boldsymbol{\Xi}^\alpha + 3\operatorname{grad}^2\varphi^\alpha \cdot \mathbb{X}^\alpha)) \,\vdots\, \operatorname{grad}^2\boldsymbol{v} \\
& - (\mathbf{\mathbb{T}} + \sum_{\alpha=1}^{n} \operatorname{grad}\varphi^\alpha \otimes \mathbb{X}^\alpha) :: \operatorname{grad}^3\boldsymbol{v} \\
& + \sum_{\alpha=1}^{n} \Big\{ (\pi^\alpha - \varrho\mu^\alpha)\dot{\varphi}^\alpha - \boldsymbol{\xi}^\alpha \cdot (\operatorname{grad}\varphi^\alpha)^{\cdot} - \boldsymbol{\Xi}^\alpha : (\operatorname{grad}^2\varphi^\alpha)^{\cdot} - \mathbb{X}^\alpha \,\vdots\, (\operatorname{grad}^3\varphi^\alpha)^{\cdot} \\
& + \mu^\alpha s^\alpha_{\text{int}} + \boldsymbol{\jmath}^\alpha \cdot \operatorname{grad}\mu^\alpha \Big\}.
\end{aligned} \tag{338}
$$

For the free-energy density ψ given by a constitutive response function $\hat{\psi}$:

$$
\psi = \hat{\psi}(\{\varphi^\alpha, \operatorname{grad}\varphi^\alpha, \operatorname{grad}^2\varphi^\alpha, \operatorname{grad}^3\varphi^\alpha\}^n_{\alpha=1}), \tag{339}
$$

with the definitions of the Larché–Cahn derivatives (262) and (268), and considering the incompressibility decomposition of the stress, hyperstress, and supra hyperstress (315), the pointwise free-energy imbalance (338) can be expressed as

$$
\begin{aligned}
0 \geq\ & - (\mathbf{T}_0 + \sum_{\alpha=1}^{n} (\operatorname{grad}\varphi^\alpha \otimes \boldsymbol{\xi}^\alpha + 2(\operatorname{grad}^2\varphi^\alpha)\boldsymbol{\Xi}^\alpha + 3\operatorname{grad}^3\varphi^\alpha : \mathbb{X}^\alpha)) : \operatorname{grad}\boldsymbol{v} \\
& - (\mathbb{T}_0 + \sum_{\alpha=1}^{n} (\operatorname{grad}\varphi^\alpha \otimes \boldsymbol{\Xi}^\alpha + 3\operatorname{grad}^2\varphi^\alpha \cdot \mathbb{X}^\alpha)) \,\vdots\, \operatorname{grad}^2\boldsymbol{v} \\
& - (\mathbf{\mathbb{T}}_0 + \sum_{\alpha=1}^{n} \operatorname{grad}\varphi^\alpha \otimes \mathbb{X}^\alpha) :: \operatorname{grad}^3\boldsymbol{v} \\
& + \sum_{\alpha=1}^{n} \Big\{ (\varrho\partial^{(\varsigma)}_{\varphi^\alpha}\hat{\psi} + \pi^\alpha - \varrho\mu^\alpha)\dot{\varphi}^\alpha + (\varrho\partial^{(\varsigma)}_{\operatorname{grad}\varphi^\alpha}\hat{\psi} - \boldsymbol{\xi}^\alpha) \cdot (\operatorname{grad}\varphi^\alpha)^{\cdot} \\
& + (\varrho\partial^{(\varsigma)}_{\operatorname{grad}^2\varphi^\alpha}\hat{\psi} - \boldsymbol{\Xi}^\alpha) : (\operatorname{grad}^2\varphi^\alpha)^{\cdot} + (\varrho\partial^{(\varsigma)}_{\operatorname{grad}^3\varphi^\alpha}\hat{\psi} - \mathbb{X}^\alpha) \,\vdots\, (\operatorname{grad}^3\varphi^\alpha)^{\cdot} \\
& + \mu^\alpha s^\alpha_{\text{int}} + \boldsymbol{\jmath}^\alpha \cdot \operatorname{grad}\mu^\alpha \Big\}.
\end{aligned} \tag{340}
$$

Adding s^α_{int} and $\boldsymbol{\jmath}^\alpha$ to the lists of dependent constitutive variables previously considered, we find that the local inequality (340) is satisfied in all processes if and only if:

- The αth microstress $\boldsymbol{\xi}^\alpha$, αth hypermicrostress $\boldsymbol{\Xi}^\alpha$, and αth supra hypermicrostress $\mathbb{X}^\alpha$ are given by constitutive response functions $\hat{\boldsymbol{\xi}}^\alpha_\varsigma$, $\hat{\boldsymbol{\Xi}}^\alpha_\varsigma$, and $\hat{\mathbb{X}}^\alpha_\varsigma$ that derive from the response function $\hat{\psi}$:

$$\left\{\begin{aligned} \boldsymbol{\xi}^\alpha &:= \hat{\boldsymbol{\xi}}^\alpha_\varsigma(\{\varphi^\alpha, \operatorname{grad}\varphi^\alpha, \operatorname{grad}^2\varphi^\alpha, \operatorname{grad}^3\varphi^\alpha\}^n_{\alpha=1}),\\ &= \varrho\partial^{(\varsigma)}_{\operatorname{grad}\varphi^\alpha}\hat{\psi},\\ \boldsymbol{\Xi}^\alpha &:= \hat{\boldsymbol{\Xi}}^\alpha_\varsigma(\{\varphi^\alpha, \operatorname{grad}\varphi^\alpha, \operatorname{grad}^2\varphi^\alpha, \operatorname{grad}^3\varphi^\alpha\}^n_{\alpha=1}),\\ &= \varrho\partial^{(\varsigma)}_{\operatorname{grad}^2\varphi^\alpha}\hat{\psi},\\ \mathbb{X}^\alpha &:= \hat{\mathbb{X}}^\alpha_\varsigma(\{\varphi^\alpha, \operatorname{grad}\varphi^\alpha, \operatorname{grad}^2\varphi^\alpha, \operatorname{grad}^3\varphi^\alpha\}^n_{\alpha=1}),\\ &= \varrho\partial^{(\varsigma)}_{\operatorname{grad}^3\varphi^\alpha}\hat{\psi}. \end{aligned}\right. \tag{341}$$

- The αth internal microforce π^α is given by a constitutive response function $\hat{\pi}^\alpha$ that differs from the αth chemical potential μ^α by a contribution derived from the response function $\hat{\psi}$:

$$\begin{aligned} \pi^\alpha &:= \hat{\pi}^\alpha_\varsigma(\{\varphi^\alpha, \operatorname{grad}\varphi^\alpha, \operatorname{grad}^2\varphi^\alpha, \operatorname{grad}^3\varphi^\alpha, \mu^\alpha\}^n_{\alpha=1}),\\ &= \varrho(\mu^\alpha_\varsigma - \partial^{(\varsigma)}_{\varphi^\alpha}\hat{\psi}). \end{aligned} \tag{342}$$

- The αth internal rate of species production s^α_{int} is given by a function of the chemical affinity and reaction rate. In addition, s^α_{int} is inversely proportional to the αth chemical potential μ^α_ς. For a detailed description, the interested reader is referred to Gurtin & Vargas [58, Remark 4.2].
- The stress $\mathbf{T}_0$, the hyperstress $\mathbb{T}_0$, and the supra hyperstress $\mathbf{T}_0$, are, respectively, given by

$$\mathbf{S} = \mathbf{T}_0 + \sum_{\alpha=1}^n(\operatorname{grad}\varphi^\alpha \otimes \boldsymbol{\xi}^\alpha + 2(\operatorname{grad}^2\varphi^\alpha)\boldsymbol{\Xi}^\alpha + 3\operatorname{grad}^3\varphi^\alpha : \mathbb{X}^\alpha), \tag{343}$$

$$\mathbb{S} = \mathbb{T}_0 + \sum_{\alpha=1}^n(\operatorname{grad}\varphi^\alpha \otimes \boldsymbol{\Xi}^\alpha + 3\operatorname{grad}^2\varphi^\alpha \cdot \mathbb{X}^\alpha), \tag{344}$$

and

$$\mathbf{S} = \mathbf{T}_0 + \sum_{\alpha=1}^n \operatorname{grad}\varphi^\alpha \otimes \mathbb{X}^\alpha. \tag{345}$$

with $\mathbf{S} := \hat{\mathbf{S}}(\operatorname{grad}\boldsymbol{v})$, $\mathbb{S} := \hat{\mathbb{S}}(\operatorname{grad}^2\boldsymbol{v})$, and $\boldsymbol{\mathsf{S}} := \hat{\boldsymbol{\mathsf{S}}}(\operatorname{grad}^3\boldsymbol{v})$. It only remains to define $\hat{\mathbf{S}}$, $\hat{\mathbb{S}}$, and $\hat{\boldsymbol{\mathsf{S}}}$ consistently with (340). We may arrive at (323), (324), and (325) again given the appropriate choices.

- The αth species flux $\jmath^\alpha$ depends smoothly on the gradient $\operatorname{grad}\mu^\alpha$ of the chemical potential μ^α for all $1 \le \alpha \le n$, and is given by a constitutive response function $\hat{\jmath}^\alpha$ of the form

$$\begin{aligned}\jmath^\alpha &:= \hat{\jmath}^\alpha(\{\varphi^\alpha, \operatorname{grad}\varphi^\alpha, \operatorname{grad}^2\varphi^\alpha, \operatorname{grad}^3\varphi^\alpha, \mu^\alpha, \operatorname{grad}\mu^\alpha\}_{\alpha=1}^n)\\ &= -\sum_{\beta=1}^n \mathbf{M}^{\alpha\beta}(\{\varphi^\alpha, \operatorname{grad}\varphi^\alpha, \operatorname{grad}^2\varphi^\alpha, \operatorname{grad}^3\varphi^\alpha, \mu^\alpha, \operatorname{grad}\mu^\alpha\}_{\alpha=1}^n)[\operatorname{grad}\mu_\varsigma^\beta],\end{aligned} \tag{346}$$

where the mobility tensor $\mathbf{M}^{\alpha\beta}$ must obey the residual dissipation inequality

$$\operatorname{grad}\mu_\varsigma^\alpha \cdot \mathbf{M}^{\alpha\beta}[\operatorname{grad}\mu_\varsigma^\beta] \ge 0. \tag{347}$$

In contrast to the theory previously developed for phase fields that are not conserved species, a complete description of the response of a material belonging to the present class consists of providing, in addition to the Clebsch–Gordan coefficients for the stresses, a scalar-valued response function $\hat{\psi}$ and a tensor valued response function $\mathbf{M}^{\alpha\beta}$ for all $1 \le \alpha, \beta \le n$. While $\hat{\psi}$ depends only on φ^α, $\operatorname{grad}\varphi^\alpha$, $\operatorname{grad}^2\varphi^\alpha$, and $\operatorname{grad}^3\varphi^\alpha$, $\mathbf{M}^{\alpha\beta}$ may also depend on μ^α and $\operatorname{grad}\mu^\alpha$. Moreover, $\mathbf{M}^{\alpha\beta}$ must satisfy the residual dissipation inequality (347) for all choices of φ^α, $\operatorname{grad}\varphi^\alpha$, $\operatorname{grad}^2\varphi^\alpha$, $\operatorname{grad}^3\varphi^\alpha$, μ^α and $\operatorname{grad}\mu^\alpha$. Also, note that expressions (323), (324), (325), (327), (328), and (329) from the nonconserved theory remain the same.

Importantly, using (341) and (342) in the field equations (121) generates the following expression

$$\varrho\mu_\varsigma^\alpha = -\operatorname{div}\left(\varrho\partial^{(\varsigma)}_{\operatorname{grad}\varphi^\alpha}\hat{\psi} - \operatorname{div}\left(\varrho\partial^{(\varsigma)}_{\operatorname{grad}^2\varphi^\alpha}\hat{\psi} - \operatorname{div}\,(\varrho\partial^{(\varsigma)}_{\operatorname{grad}^3\varphi^\alpha}\hat{\psi})\right)\right) + \varrho\partial^{(\varsigma)}_{\varphi^\alpha}\hat{\psi} - \gamma_\varsigma^\alpha, \tag{348}$$

for the chemical potential which, in conjunction with the constitutive relation (346) for the species flux and the pointwise species balance (335), yields the conserved field equations

$$\varrho\dot{\varphi}^\alpha = \operatorname{div}\Big(\sum_{\beta=1}^\alpha \mathbf{M}^{\alpha\beta}[\operatorname{grad}\mu_\varsigma^\beta]\Big) + s^\alpha, \tag{349}$$

for the αth phase field. Owing to the dependence of the response function $\hat{\psi}$ on $\operatorname{grad}^3\varphi$, μ_ς^α as determined by (348) involves sixth-order spatial derivatives of φ^α and (349) thus includes eighth-order spatial derivatives of φ^α.

In analogy to the comment immediately after expression (330), this suggests the possibility of referring to (349) with μ_ς^α given by (348) as a 'conserved third-grade phase-field equation'.

Synopsis, chapter 6

Constitutive relations

Free-energy density

$$\psi = f(\{\varphi^\alpha\}_{\alpha=1}^n) + \tfrac{1}{2}\sum_{\alpha=1}^{n}(\|\boldsymbol{y}^\alpha\|^2 + \|\boldsymbol{\Upsilon}^\alpha\|^2 + \|\mathbb{Y}^\alpha\|^2)$$

where f is a potential function, $\boldsymbol{y}^\alpha := \operatorname{grad}\varphi^\alpha$, $\boldsymbol{\Upsilon}^\alpha := \operatorname{grad}^2\varphi^\alpha$, $\mathbb{Y}^\alpha := \operatorname{grad}^3\varphi^\alpha$, $\mathbf{B}$:=sym
($\operatorname{grad}\boldsymbol{v}$), $\mathbb{B} := \operatorname{grad}^2\boldsymbol{v}$, and $\mathbf{B} := \operatorname{grad}^3\boldsymbol{v}$

Microkinetics

(i) αth microstress

$$(\boldsymbol{\xi}^\alpha)_i = \varrho\ell^\alpha_{(2,1)}(\boldsymbol{y}^\alpha)_i$$

(ii) αth hypermicrostress

$$(\boldsymbol{\Xi}^\alpha)_{ij} = \varrho\ell^\alpha_{(4,1)}\delta_{ij}(\boldsymbol{\Upsilon}^\alpha)_{kk} + 2\varrho\ell^\alpha_{(4,2)}(\boldsymbol{\Upsilon}^\alpha)_{ij}$$

(iii) αth supra hypermicrostress

$$\begin{aligned}(\mathbb{X}^\alpha)_{ijk} &= 3\varrho\ell^\alpha_{(6,1)}(\delta_{ij}(\mathbb{Y}^\alpha)_{kmm} + \delta_{ik}(\mathbb{Y}^\alpha)_{jmm} + \delta_{jk}(\mathbb{Y}^\alpha)_{imm})\\ &\quad + 6\varrho\ell^\alpha_{(6,2)}(\mathbb{Y}^\alpha)_{ijk}\end{aligned}$$

Kinetics

(i) Stress

$$\mathbf{T} = \mathbf{S} - \mathbf{1}g - \sum_{\alpha=1}^{n}(\operatorname{grad}\varphi^\alpha \otimes \boldsymbol{\xi}^\alpha + 2(\operatorname{grad}^2\varphi^\alpha)\boldsymbol{\Xi}^\alpha + 3\operatorname{grad}^3\varphi^\alpha : \mathbb{X}^\alpha)$$

with

$$(\mathbf{S})_{ij} = 2\ell_{(4,1)}(\mathbf{B})_{ij}$$

(ii) Hyperstress

$$\mathbb{T} = \mathbb{S} - \mathbf{1}\otimes\boldsymbol{g} - \sum_{\alpha=1}^{n}(\operatorname{grad}\varphi^\alpha \otimes \boldsymbol{\Xi}^\alpha + 3\operatorname{grad}^2\varphi^\alpha \cdot \mathbb{X}^\alpha)$$

with

$$(\mathbb{S})_{ijk} = 2\ell_{(6,1)}((\mathbb{B})_{kij} + (\mathbb{B})_{jik} - \tfrac{1}{2}(\delta_{ij}(\mathbb{B})_{kmm} + \delta_{ik}(\mathbb{B})_{jmm}))$$
$$+ 2\ell_{(6,2)}(\mathbb{B})_{ijk} + \ell_{(6,3)}(\delta_{jk}(\mathbb{B})_{imm} - \tfrac{1}{2}(\delta_{ij}(\mathbb{B})_{kmm} + \delta_{ik}(\mathbb{B})_{jmm}))$$

(iii) Second hyperstress

$$\mathbb{T} = \mathbf{S} - \mathbf{1} \otimes \mathbf{G} - \sum_{\alpha=1}^{n} \operatorname{grad} \varphi^\alpha \otimes \mathbb{X}^\alpha$$

with

$$\begin{aligned}(\mathbf{S})_{ijkl} = 3\ell_{(8,1)}((&\delta_{ij}(\mathbf{B})_{lknn} + \delta_{ij}(\mathbf{B})_{klnn} + \delta_{ik}(\mathbf{B})_{jlnn} + \delta_{ik}(\mathbf{B})_{ljnn} \\ &+ \delta_{il}(\mathbf{B})_{jknn} + \delta_{il}(\mathbf{B})_{kjnn}) - 5((\mathbf{B})_{lijk} + (\mathbf{B})_{jikl} + (\mathbf{B})_{kijl})) \\ &+ 3\ell_{(8,2)}(\delta_{jk}(\mathbf{B})_{limm} + \delta_{jl}(\mathbf{B})_{kimm} + \delta_{kl}(\mathbf{B})_{jimm} \\ &- ((\mathbf{B})_{lijk} + (\mathbf{B})_{jikl} + (\mathbf{B})_{kijl})) \\ &+ 3\ell_{(8,3)}(\delta_{kl}(\mathbf{B})_{ijnn} + \delta_{jl}(\mathbf{B})_{ikmm} + \delta_{jk}(\mathbf{B})_{ilmm} \\ &- ((\mathbf{B})_{lijk} + (\mathbf{B})_{jikl} + (\mathbf{B})_{kijl})) \\ &+ 6\ell_{(8,4)}(\mathbf{B})_{ijkl}.\end{aligned}$$

Field equations for the phase fields

(i) Nonconserved case

$$-\pi^\alpha_{\text{dis}} = \operatorname{div}\Big(\varrho\partial_{\operatorname{grad}\varphi^\alpha}\hat{\psi} - \operatorname{div}\big(\varrho\partial_{\operatorname{grad}^2\varphi^\alpha}\hat{\psi} - \operatorname{div}(\varrho\partial_{\operatorname{grad}^3\varphi^\alpha}\hat{\psi})\big)\Big)$$
$$- \partial_{\varphi^\alpha}\hat{\psi} + \gamma^\alpha$$

(ii) Conserved case

$$\varrho\mu^\alpha_\varsigma = -\operatorname{div}\Big(\varrho\partial^{(\varsigma)}_{\operatorname{grad}\varphi^\alpha}\hat{\psi} - \operatorname{div}\big(\varrho\partial^{(\varsigma)}_{\operatorname{grad}^2\varphi^\alpha}\hat{\psi} - \operatorname{div}(\varrho\partial^{(\varsigma)}_{\operatorname{grad}^3\varphi^\alpha}\hat{\psi})\big)\Big)$$
$$+ \partial^{(\varsigma)}_{\varphi^\alpha}\hat{\psi} - \gamma^\alpha_\varsigma$$

$$\varrho\dot{\varphi}^\alpha = \operatorname{div}\Big(\sum_{\beta=1}^{\alpha} \mathbf{M}^{\alpha\beta}[\operatorname{grad}\mu^\beta_\varsigma]\Big) + s^\alpha$$

Part IV
Environmental Conditions and Boundary Conditions

Chapter 7
Environmental Surface Balances and Imbalances

In this chapter, we capitalize on the Fried & Gurtin's [20, 60] procedure to determine thermodynamically consistent boundary conditions by tailoring the surface balances of forces, microforces, torques, and microtorques as well as the surface power balances as similarly proposed by Espath & Calo [32]. Analogous boundary conditions to those presented by Fried & Gurtin [20] are also provided by Espath et al. [33] and Duda et al. [61].

1 Surface Balances

In taking the surface $\mathcal{S}$ in expressions (205), (206), (207), and (208) to the limit such that the surface coincides with the body's boundary, $\mathcal{S}_{\text{env}} \subseteq \partial\mathcal{B}$, we have that the

$$\int\limits_{\mathcal{S}_{\text{env}}} (\xi^{\alpha}_{\mathcal{S}\,\text{env}} + \xi^{\alpha}_{\mathcal{S}^*})\,\mathrm{d}a + \int\limits_{\partial\mathcal{S}_{\text{env}}} \xi^{\alpha}_{\partial\mathcal{S}\,\text{env}}\,\mathrm{d}\sigma + \int\limits_{\mathcal{C}_{\text{env}}} \xi^{\alpha}_{\mathcal{C}\,\text{env}}\,\mathrm{d}\sigma + \xi^{\alpha}_{\partial^2\mathcal{S}\,\text{env}} + \xi^{\alpha}_{\partial\mathcal{C}\,\text{env}} + \xi^{\alpha}_{\mathcal{O}\,\text{env}} \overset{!}{=} 0, \tag{350}$$

$$\forall\, \mathcal{S}_{\text{env}} \subset \partial\mathcal{B} \quad \text{and} \quad \forall\, t, \quad \text{with} \quad 1 \leq \alpha \leq n,$$

$$\int\limits_{\mathcal{S}_{\text{env}}} (\boldsymbol{t}_{\mathcal{S}\,\text{env}} + \boldsymbol{t}_{\mathcal{S}^*})\,\mathrm{d}a + \int\limits_{\partial\mathcal{S}_{\text{env}}} \boldsymbol{t}_{\partial\mathcal{S}\,\text{env}}\,\mathrm{d}\sigma + \int\limits_{\mathcal{C}_{\text{env}}} \boldsymbol{t}_{\mathcal{C}\,\text{env}}\,\mathrm{d}\sigma + \boldsymbol{t}_{\partial^2\mathcal{S}\,\text{env}} + \boldsymbol{t}_{\partial\mathcal{C}\,\text{env}} + \boldsymbol{t}_{\mathcal{O}\,\text{env}} \overset{!}{=} \mathbf{0}, \tag{351}$$

$$\forall\, \mathcal{S}_{\text{env}} \subset \partial\mathcal{B} \quad \text{and} \quad \forall\, t,$$

L. Espath, *Mechanics and Geometry of Enriched Continua*,
https://doi.org/10.1007/978-3-031-28934-7_7

$$\int_{\mathcal{S}_{\text{env}}} (\varpi^{\alpha}_{\mathcal{S}\,\text{env}} + \varpi^{\alpha}_{\mathcal{S}^*})\,\mathrm{d}a + \int_{\mathcal{S}_{\text{env}}} \boldsymbol{r}(\xi^{\alpha}_{\mathcal{S}\,\text{env}} + \xi^{\alpha}_{\mathcal{S}^*})\,\mathrm{d}a$$
$$+ \int_{\partial\mathcal{S}_{\text{env}}} \varpi^{\alpha}_{\partial\mathcal{S}\,\text{env}}\,\mathrm{d}\sigma + \int_{\partial\mathcal{S}_{\text{env}}} \boldsymbol{r}\xi^{\alpha}_{\partial\mathcal{S}\,\text{env}}\,\mathrm{d}\sigma + \int_{\mathcal{C}_{\text{env}}} \varpi^{\alpha}_{\mathcal{C}\,\text{env}}\,\mathrm{d}\sigma + \int_{\mathcal{C}_{\text{env}}} \boldsymbol{r}\xi^{\alpha}_{\mathcal{C}\,\text{env}}\,\mathrm{d}\sigma$$
$$+ \boldsymbol{r}_{\partial^2\mathcal{S}\,\text{env}}\xi^{\alpha}_{\partial^2\mathcal{S}\,\text{env}} + \boldsymbol{r}_{\partial\mathcal{C}\,\text{env}}\xi^{\alpha}_{\partial\mathcal{C}\,\text{env}} + \boldsymbol{r}_{\mathcal{O}\,\text{env}}\xi^{\alpha}_{\mathcal{O}\,\text{env}} \stackrel{!}{=} \mathbf{0}, \tag{352}$$

$$\forall\,\mathcal{S}_{\text{env}} \subset \partial\mathcal{B} \quad \text{and} \quad \forall t, \quad \text{with} \quad 1 \le \alpha \le n,$$

and

$$\int_{\mathcal{S}_{\text{env}}} (\boldsymbol{m}_{\mathcal{S}\,\text{env}} + \boldsymbol{m}_{\mathcal{S}^*})\,\mathrm{d}a + \int_{\mathcal{S}_{\text{env}}} \boldsymbol{r} \times (\boldsymbol{t}_{\mathcal{S}\,\text{env}} + \boldsymbol{t}_{\mathcal{S}^*})\,\mathrm{d}a$$
$$+ \int_{\partial\mathcal{S}_{\text{env}}} \boldsymbol{m}_{\partial\mathcal{S}\,\text{env}}\,\mathrm{d}\sigma + \int_{\partial\mathcal{S}_{\text{env}}} \boldsymbol{r} \times \boldsymbol{t}_{\partial\mathcal{S}\,\text{env}}\,\mathrm{d}\sigma + \int_{\mathcal{C}_{\text{env}}} \boldsymbol{m}_{\mathcal{C}\,\text{env}}\,\mathrm{d}\sigma + \int_{\mathcal{C}_{\text{env}}} \boldsymbol{r} \times \boldsymbol{t}_{\mathcal{C}\,\text{env}}\,\mathrm{d}\sigma$$
$$+ \boldsymbol{r}_{\partial^2\mathcal{S}\,\text{env}} \times \boldsymbol{t}_{\partial^2\mathcal{S}\,\text{env}} + \boldsymbol{r}_{\partial\mathcal{C}\,\text{env}} \times \boldsymbol{t}_{\partial\mathcal{C}\,\text{env}} + \boldsymbol{r}_{\mathcal{O}\,\text{env}} \times \boldsymbol{t}_{\mathcal{O}\,\text{env}} \stackrel{!}{=} \mathbf{0}, \tag{353}$$

$$\forall\,\mathcal{S}_{\text{env}} \subset \partial\mathcal{B} \quad \text{and} \quad \forall t.$$

Naturally, the subscript *env* makes reference to environmental, for instance, $\xi^{\alpha}_{\mathcal{S}\,\text{env}}$ and $\boldsymbol{t}_{\mathcal{S}\text{env}}$ are, respectively, the surface microtraction and surface traction resulting from the action of the environment on $\mathcal{S}_{\text{env}}$. Also, we refer to these expressions above as environmental surface balances. Note that the balances above are set to zero. Thus, no constitutive equations for the interface are allowed through additional microtraction- and traction-like fields and couple-fields. However, it is possible to incorporate the action of tangential microstresses and stresses, tangential hypermicrostresses and hyperstresses, and tangential supra hypermicrostresses and hyperstresses.

Next, consider the case where the phase fields represent the concentration of conserved species. Thus, in addition to the surface balances of microtractions (350), forces (351), microtorques (352), and torques (353), we supplement the system with the environmental surface species balance on $\overline{\mathcal{S}}$. This balance results from the balance between the environmental mass transfer rate and the mass transfer from the body across the surface $-\overline{\mathcal{S}}$ and reads

$$\int_{\overline{\mathcal{S}}} (\jmath_{\text{env}} + \boldsymbol{j} \cdot \boldsymbol{n}) \, \mathrm{d}a = 0, \tag{354}$$

which by localization renders

$$\jmath_{\text{env}} + \boldsymbol{j} \cdot \boldsymbol{n} = 0. \tag{355}$$

Here, $\jmath_{\text{env}}$ represents the transfer of mass from the environment onto $\overline{\mathcal{S}}$.

2 Surface Imbalances

Similarly, expressions (225) and (227) may be specialized such that the surface coincides with the boundary. Thus, the actual surface external power $\mathcal{W}^{\flat}_{\text{ext}}$ is given by

$$\begin{aligned}
&\mathcal{W}^{\flat}_{\text{ext}}(\overline{\mathcal{S}}_{\text{env}}; \{\dot{\varphi}^{\alpha}_{\text{env}}\}^{n}_{\alpha=1}) := \\
&\quad \sum_{\alpha=1}^{n} \Bigg\{ \int_{\overline{\mathcal{S}}_{\text{env}}} (\xi^{\alpha}_{\mathcal{S}\,\text{env}} \dot{\varphi}^{\alpha}_{\text{env}} + {}^{2}\xi^{\alpha}_{\mathcal{S}\,\text{env}} \partial_n \dot{\varphi}^{\alpha}_{\text{env}} + {}^{3}\xi^{\alpha}_{\mathcal{S}\,\text{env}} \partial^{2}_{n} \dot{\varphi}^{\alpha}_{\text{env}}) \, \mathrm{d}a \\
&\qquad + \int_{\overline{\partial\mathcal{S}}_{\text{env}}} (\xi^{\alpha}_{\partial\mathcal{S}\,\text{env}} \dot{\varphi}^{\alpha}_{\text{env}} + {}^{2}_{n}\xi^{\alpha}_{\partial\mathcal{S}\,\text{env}} \partial_n \dot{\varphi}^{\alpha}_{\text{env}} + {}^{2}_{\nu}\xi^{\alpha}_{\partial\mathcal{S}\,\text{env}} \partial_{\nu} \dot{\varphi}^{\alpha}_{\text{env}}) \, \mathrm{d}\sigma \\
&\qquad + \int_{\overline{\mathcal{C}}_{\text{env}}} (\xi^{\alpha}_{\mathcal{C}\,\text{env}} \dot{\varphi}^{\alpha}_{\text{env}} + {}^{2}_{n}\xi^{\alpha}_{\mathcal{C}\,\text{env}} \partial_n \dot{\varphi}^{\alpha}_{\text{env}} + {}^{2}_{\nu}\xi^{\alpha}_{\mathcal{C}\,\text{env}} \partial_{\nu} \dot{\varphi}^{\alpha}_{\text{env}}) \, \mathrm{d}\sigma \\
&\qquad + \xi^{\alpha}_{\partial^2\mathcal{S}\,\text{env}} \dot{\varphi}^{\alpha}_{\text{env}}|_{\partial^2\mathcal{S}_{\text{env}}} + \xi^{\alpha}_{\partial\mathcal{C}\,\text{env}} \dot{\varphi}^{\alpha}_{\text{env}}|_{\partial\mathcal{C}_{\text{env}}} + \xi^{\alpha}_{\mathcal{O}\text{env}} \dot{\varphi}^{\alpha}_{\text{env}}|_{\mathcal{O}_{\text{env}}} \Bigg\},
\end{aligned} \tag{356}$$

and the actual surface external power $\mathcal{W}^{\sharp}_{\text{ext}}$ by

$$
\begin{aligned}
&\mathcal{W}^{\sharp}_{\text{ext}}(\overline{\mathcal{S}}_{\text{env}}; \boldsymbol{v}_{\text{env}}) := \\
&\qquad \int\limits_{\overline{\mathcal{S}}_{\text{env}}} \big(\boldsymbol{t}_{\mathcal{S}\text{env}} \cdot \boldsymbol{v}_{\text{env}} + {}^{2}\boldsymbol{t}_{\mathcal{S}\text{env}} \cdot \partial_n \boldsymbol{v}_{\text{env}} + {}^{3}\boldsymbol{t}_{\mathcal{S}\text{env}} \cdot \partial_n^2 \boldsymbol{v}_{\text{env}}\big)\, \mathrm{d}a \\
&\qquad + \int\limits_{\partial\overline{\mathcal{S}}_{\text{env}}} \big(\boldsymbol{t}_{\partial\mathcal{S}\text{env}} \cdot \boldsymbol{v}_{\text{env}} + {}^{2}_{n}\boldsymbol{t}_{\partial\mathcal{S}\text{env}} \cdot \partial_n \boldsymbol{v}_{\text{env}} + {}^{2}_{\nu}\boldsymbol{t}_{\partial\mathcal{S}\text{env}} \cdot \partial_\nu \boldsymbol{v}_{\text{env}}\big)\, \mathrm{d}\sigma \\
&\qquad + \int\limits_{\overline{\mathcal{C}}_{\text{env}}} \big(\boldsymbol{t}_{\mathcal{C}\text{env}} \cdot \boldsymbol{v}_{\text{env}} + {}^{2}_{n}\boldsymbol{t}_{\mathcal{C}\text{env}} \cdot \partial_n \boldsymbol{v}_{\text{env}} + {}^{2}_{\nu}\boldsymbol{t}_{\mathcal{C}\text{env}} \cdot \partial_\nu \boldsymbol{v}_{\text{env}}\big)\, \mathrm{d}\sigma \\
&\qquad + \boldsymbol{t}_{\partial^2\mathcal{S}\text{env}} \cdot \boldsymbol{v}_{\text{env}}|_{\partial^2\mathcal{S}_{\text{env}}} + \boldsymbol{t}_{\partial\mathcal{C}\text{env}} \cdot \boldsymbol{v}_{\text{env}}|_{\partial\mathcal{C}_{\text{env}}} + \boldsymbol{t}_{\mathcal{O}\text{env}} \cdot \boldsymbol{v}_{\text{env}}|_{\mathcal{O}_{\text{env}}}.
\end{aligned}
\tag{357}
$$

The internal surface power remains related to the internal interactions of the body $\mathcal{B}$, that is, $\mathcal{W}^{\flat}_{\text{int}} := \mathcal{W}^{\flat}_{\text{int}}(\overline{\mathcal{S}}_{\text{env}}; \{\dot{\varphi}^{\alpha}\}_{\alpha=1}^{n})$ and $\mathcal{W}^{\sharp}_{\text{int}} := \mathcal{W}^{\sharp}_{\text{int}}(\overline{\mathcal{S}}_{\text{env}}; \boldsymbol{v})$. Thus, the surface power balances (222) and (223) become

$$
\mathcal{W}^{\flat}_{\text{int}}(\overline{\mathcal{S}}_{\text{env}}; \{\dot{\varphi}^{\alpha}\}_{\alpha=1}^{n}) = \mathcal{W}^{\flat}_{\text{ext}}(\overline{\mathcal{S}}_{\text{env}}; \{\dot{\varphi}^{\alpha}_{\text{env}}\}_{\alpha=1}^{n}), \tag{358}
$$

and

$$
\mathcal{W}^{\sharp}_{\text{int}}(\overline{\mathcal{S}}_{\text{env}}; \boldsymbol{v}) = \mathcal{W}^{\sharp}_{\text{ext}}(\overline{\mathcal{S}}_{\text{env}}; \boldsymbol{v}_{\text{env}}), \tag{359}
$$

respectively.

Next, note that, given the surface power balances (358) and (359), we stipulate that

$$
\begin{cases}
\xi^{\alpha}_{\mathcal{S}\text{env}} = \xi^{\alpha}_{\mathcal{S}}, & {}^{2}\xi^{\alpha}_{\mathcal{S}\text{env}} = {}^{2}\xi^{\alpha}_{\mathcal{S}}, \qquad {}^{3}\xi^{\alpha}_{\mathcal{S}\text{env}} = {}^{3}\xi^{\alpha}_{\mathcal{S}}, \\
\xi^{\alpha}_{\partial\mathcal{S}\text{env}} = \xi^{\alpha}_{\partial\mathcal{S}}, & {}^{2}_{n}\xi^{\alpha}_{\partial\mathcal{S}\text{env}} = {}^{2}_{n}\xi^{\alpha}_{\partial\mathcal{S}}, \qquad {}^{2}_{\nu}\xi^{\alpha}_{\partial\mathcal{S}\text{env}} = {}^{2}_{\nu}\xi^{\alpha}_{\partial\mathcal{S}}, \\
\xi^{\alpha}_{\mathcal{C}\text{env}} = \xi^{\alpha}_{\mathcal{C}}, & {}^{2}_{n}\xi^{\alpha}_{\mathcal{C}\text{env}} = {}^{2}_{n}\xi^{\alpha}_{\mathcal{C}}, \qquad {}^{2}_{\nu}\xi^{\alpha}_{\mathcal{C}\text{env}} = {}^{2}_{\nu}\xi^{\alpha}_{\mathcal{C}}, \\
\xi^{\alpha}_{\partial^2\mathcal{S}\text{env}} = \xi^{\alpha}_{\partial^2\mathcal{S}}, & \xi^{\alpha}_{\partial\mathcal{C}\text{env}} = \xi^{\alpha}_{\partial\mathcal{C}}, \qquad \xi^{\alpha}_{\mathcal{O}\text{env}} = \xi^{\alpha}_{\mathcal{O}},
\end{cases}
\tag{360}
$$

and

$$
\begin{cases}
\boldsymbol{t}_{\mathcal{S}\text{env}} = \boldsymbol{t}_{\mathcal{S}}, & {}^{2}\boldsymbol{t}_{\mathcal{S}\text{env}} = {}^{2}\boldsymbol{t}_{\mathcal{S}}, \qquad \boldsymbol{t}_{\mathcal{C}\text{env}} = \boldsymbol{t}_{\mathcal{C}}, \\
\boldsymbol{t}_{\partial\mathcal{S}\text{env}} = \boldsymbol{t}_{\partial\mathcal{S}}, & {}^{2}_{n}\boldsymbol{t}_{\partial\mathcal{S}\text{env}} = {}^{2}_{n}\boldsymbol{t}_{\partial\mathcal{S}}, \qquad {}^{2}_{\nu}\boldsymbol{t}_{\partial\mathcal{S}\text{env}} = {}^{2}_{\nu}\boldsymbol{t}_{\partial\mathcal{S}}, \\
\boldsymbol{t}_{\mathcal{C}\text{env}} = \boldsymbol{t}_{\mathcal{C}}, & {}^{2}_{n}\boldsymbol{t}_{\mathcal{C}\text{env}} = {}^{2}_{n}\boldsymbol{t}_{\mathcal{C}}, \qquad {}^{2}_{\nu}\boldsymbol{t}_{\mathcal{C}\text{env}} = {}^{2}_{\nu}\boldsymbol{t}_{\mathcal{C}}, \\
\boldsymbol{t}_{\partial^2\mathcal{S}\text{env}} = \boldsymbol{t}_{\partial^2\mathcal{S}}, & \boldsymbol{t}_{\partial\mathcal{C}\text{env}} = \boldsymbol{t}_{\partial\mathcal{C}}, \qquad \boldsymbol{t}_{\mathcal{O}\text{env}} = \boldsymbol{t}_{\mathcal{O}}.
\end{cases}
\tag{361}
$$

To allow for a more general type of boundary conditions, the environmental surface power balances may be relaxed to consider a dissipative environment. That is, the surface power balances (358) and (359) are transformed into environmental surface power imbalances as follows

$$\mathcal{W}^{\flat}_{\text{ext}}(\overline{\mathcal{S}}_{\text{env}}; \{\dot{\varphi}^{\alpha}_{\text{env}}\}^{n}_{\alpha=1}) - \mathcal{W}^{\flat}_{\text{int}}(\overline{\mathcal{S}}_{\text{env}}; \{\dot{\varphi}^{\alpha}\}^{n}_{\alpha=1}) =: \Lambda^{\flat}_{\text{env}}(\{\dot{\varphi}^{\alpha}, \dot{\varphi}^{\alpha}_{\text{env}}\}^{n}_{\alpha=1}) \geq 0, \quad (362)$$

and

$$\mathcal{W}^{\sharp}_{\text{ext}}(\overline{\mathcal{S}}_{\text{env}}; \boldsymbol{v}_{\text{env}}) - \mathcal{W}^{\sharp}_{\text{int}}(\overline{\mathcal{S}}_{\text{env}}; \boldsymbol{v}) =: \Lambda^{\sharp}_{\text{env}}(\boldsymbol{v}, \boldsymbol{v}_{\text{env}}) \geq 0. \quad (363)$$

Moreover, given the general surface power balances (222) and (223), we have the following explicit form for the environmental surface power imbalances

$$\begin{aligned}
&\Lambda^{\flat}_{\text{env}}(\{\dot{\varphi}^{\alpha}, \dot{\varphi}^{\alpha}_{\text{env}}\}^{n}_{\alpha=1}) = \\
&\quad \sum_{\alpha=1}^{n} \Bigg\{ \int\limits_{\overline{\mathcal{S}}_{\text{env}}} \big(\xi^{\alpha}_{\mathcal{S}\,\text{env}} \dot{\varphi}^{\alpha}_{\text{env}} + {}^{2}\xi^{\alpha}_{\mathcal{S}\,\text{env}} \partial_n \dot{\varphi}^{\alpha}_{\text{env}} + {}^{3}\xi^{\alpha}_{\mathcal{S}\,\text{env}} \partial^{2}_{n} \dot{\varphi}^{\alpha}_{\text{env}}\big)\, \mathrm{d}a \\
&\qquad - \int\limits_{\overline{\mathcal{S}}_{\text{env}}} \big(\xi^{\alpha}_{\mathcal{S}} \dot{\varphi}^{\alpha} + {}^{2}\xi^{\alpha}_{\mathcal{S}} \partial_n \dot{\varphi}^{\alpha} + {}^{3}\xi^{\alpha}_{\mathcal{S}} \partial^{2}_{n} \dot{\varphi}^{\alpha}\big)\, \mathrm{d}a \\
&\qquad + \int\limits_{\partial\overline{\mathcal{S}}_{\text{env}}} \big(\xi^{\alpha}_{\partial\mathcal{S}\,\text{env}} \dot{\varphi}^{\alpha}_{\text{env}} + {}^{2}_{n}\xi^{\alpha}_{\partial\mathcal{S}\,\text{env}} \partial_n \dot{\varphi}^{\alpha}_{\text{env}} + {}^{2}_{\nu}\xi^{\alpha}_{\partial\mathcal{S}\,\text{env}} \partial_\nu \dot{\varphi}^{\alpha}_{\text{env}}\big)\, \mathrm{d}\sigma \\
&\qquad - \int\limits_{\partial\overline{\mathcal{S}}_{\text{env}}} \big(\xi^{\alpha}_{\partial\mathcal{S}} \dot{\varphi}^{\alpha} + {}^{2}_{n}\xi^{\alpha}_{\partial\mathcal{S}} \partial_n \dot{\varphi}^{\alpha} + {}^{2}_{\nu}\xi^{\alpha}_{\partial\mathcal{S}} \partial_\nu \dot{\varphi}^{\alpha}\big)\, \mathrm{d}\sigma \\
&\qquad + \int\limits_{\overline{\mathcal{C}}_{\text{env}}} \big(\xi^{\alpha}_{\mathcal{C}\,\text{env}} \dot{\varphi}^{\alpha}_{\text{env}} + {}^{2}_{n}\xi^{\alpha}_{\mathcal{C}\,\text{env}} \partial_n \dot{\varphi}^{\alpha}_{\text{env}} + {}^{2}_{\nu}\xi^{\alpha}_{\mathcal{C}\,\text{env}} \partial_\nu \dot{\varphi}^{\alpha}_{\text{env}}\big)\, \mathrm{d}\sigma \\
&\qquad - \int\limits_{\overline{\mathcal{C}}_{\text{env}}} \big(\xi^{\alpha}_{\mathcal{C}} \dot{\varphi}^{\alpha} + {}^{2}_{n}\xi^{\alpha}_{\mathcal{C}} \partial_n \dot{\varphi}^{\alpha} + {}^{2}_{\nu}\xi^{\alpha}_{\mathcal{C}} \partial_\nu \dot{\varphi}^{\alpha}\big)\, \mathrm{d}\sigma \\
&\qquad + \xi^{\alpha}_{\partial^{2}\mathcal{S}\,\text{env}} \dot{\varphi}^{\alpha}_{\text{env}}\big|_{\partial^{2}\mathcal{S}_{\text{env}}} + \xi^{\alpha}_{\partial\mathcal{C}\,\text{env}} \dot{\varphi}^{\alpha}_{\text{env}}\big|_{\partial\mathcal{C}_{\text{env}}} + \xi^{\alpha}_{\mathcal{O}\,\text{env}} \dot{\varphi}^{\alpha}_{\text{env}}\big|_{\mathcal{O}_{\text{env}}} \\
&\qquad - \xi^{\alpha}_{\partial^{2}\mathcal{S}} \dot{\varphi}^{\alpha}\big|_{\partial^{2}\mathcal{S}_{\text{env}}} - \xi^{\alpha}_{\partial\mathcal{C}} \dot{\varphi}^{\alpha}\big|_{\partial\mathcal{C}_{\text{env}}} - \xi^{\alpha}_{\mathcal{O}} \dot{\varphi}^{\alpha}\big|_{\mathcal{O}_{\text{env}}} \Bigg\} \geq 0, \qquad (364)
\end{aligned}$$

and

$$
\begin{aligned}
\Lambda^{\sharp}_{\mathrm{env}}(\boldsymbol{v}, \boldsymbol{v}_{\mathrm{env}}) = &\\
&\int_{\overline{\mathcal{S}}_{\mathrm{env}}} \left(\boldsymbol{t}_{\mathcal{S}\mathrm{env}} \cdot \boldsymbol{v}_{\mathrm{env}} + {}^{2}\boldsymbol{t}_{\mathcal{S}\mathrm{env}} \cdot \partial_n \boldsymbol{v}_{\mathrm{env}} + {}^{3}\boldsymbol{t}_{\mathcal{S}\mathrm{env}} \cdot \partial_n^2 \boldsymbol{v}_{\mathrm{env}} \right) \mathrm{d}a \\
&- \int_{\overline{\mathcal{S}}_{\mathrm{env}}} \left(\boldsymbol{t}_{\mathcal{S}} \cdot \boldsymbol{v} + {}^{2}\boldsymbol{t}_{\mathcal{S}} \cdot \partial_n \boldsymbol{v} + {}^{3}\boldsymbol{t}_{\mathcal{S}} \cdot \partial_n^2 \boldsymbol{v} \right) \mathrm{d}a \\
&+ \int_{\overline{\partial\mathcal{S}}_{\mathrm{env}}} \left(\boldsymbol{t}_{\partial\mathcal{S}\mathrm{env}} \cdot \boldsymbol{v}_{\mathrm{env}} + {}^{2}_{n}\boldsymbol{t}_{\partial\mathcal{S}\mathrm{env}} \cdot \partial_n \boldsymbol{v}_{\mathrm{env}} + {}^{2}_{\nu}\boldsymbol{t}_{\partial\mathcal{S}\mathrm{env}} \cdot \partial_\nu \boldsymbol{v}_{\mathrm{env}} \right) \mathrm{d}\sigma \\
&- \int_{\overline{\partial\mathcal{S}}_{\mathrm{env}}} \left(\boldsymbol{t}_{\partial\mathcal{S}} \cdot \boldsymbol{v} + {}^{2}_{n}\boldsymbol{t}_{\partial\mathcal{S}} \cdot \partial_n \boldsymbol{v} + {}^{2}_{\nu}\boldsymbol{t}_{\partial\mathcal{S}} \cdot \partial_\nu \boldsymbol{v} \right) \mathrm{d}\sigma \\
&+ \int_{\overline{\mathcal{C}}_{\mathrm{env}}} \left(\boldsymbol{t}_{\mathcal{C}\mathrm{env}} \cdot \boldsymbol{v}_{\mathrm{env}} + {}^{2}_{n}\boldsymbol{t}_{\mathcal{C}\mathrm{env}} \cdot \partial_n \boldsymbol{v}_{\mathrm{env}} + {}^{2}_{\nu}\boldsymbol{t}_{\mathcal{C}\mathrm{env}} \cdot \partial_\nu \boldsymbol{v}_{\mathrm{env}} \right) \mathrm{d}\sigma \\
&- \int_{\overline{\mathcal{C}}_{\mathrm{env}}} \left(\boldsymbol{t}_{\mathcal{C}} \cdot \boldsymbol{v} + {}^{2}_{n}\boldsymbol{t}_{\mathcal{C}} \cdot \partial_n \boldsymbol{v} + {}^{2}_{\nu}\boldsymbol{t}_{\mathcal{C}} \cdot \partial_\nu \boldsymbol{v} \right) \mathrm{d}\sigma \\
&+ \boldsymbol{t}_{\partial^2\mathcal{S}\mathrm{env}} \cdot \boldsymbol{v}_{\mathrm{env}}|_{\partial^2\mathcal{S}_{\mathrm{env}}} + \boldsymbol{t}_{\partial\mathcal{C}\mathrm{env}} \cdot \boldsymbol{v}_{\mathrm{env}}|_{\partial\mathcal{C}_{\mathrm{env}}} + \boldsymbol{t}_{\mathcal{O}\mathrm{env}} \cdot \boldsymbol{v}_{\mathrm{env}}|_{\mathcal{O}_{\mathrm{env}}} \\
&- \boldsymbol{t}_{\partial^2\mathcal{S}} \cdot \boldsymbol{v}|_{\partial^2\mathcal{S}_{\mathrm{env}}} - \boldsymbol{t}_{\partial\mathcal{C}} \cdot \boldsymbol{v}|_{\partial\mathcal{C}_{\mathrm{env}}} - \boldsymbol{t}_{\mathcal{O}} \cdot \boldsymbol{v}|_{\mathcal{O}_{\mathrm{env}}} \geq 0.
\end{aligned}
\tag{365}
$$

Next, consider the case where the phase fields represent the concentration of conserved species. As we augmented the partwise free-energy imbalance (249) with the energy transfer rate to $\overline{\mathcal{P}}$ due to species transport to arrive at (333), we augment (362) with the energy transfer rate to $\overline{\mathcal{S}}$ due to species transport. We then consider the environmental energy transfer rate and the energy transfer rate from the body to the surface $\overline{\mathcal{S}}$. Thus, we add the environmental energy transfer rate to the actual surface external power

$$
\sum_{\alpha=1}^{n} \int_{\overline{\mathcal{S}}_{\mathrm{env}}} \mu^{\alpha}_{\mathrm{env}} \jmath^{\alpha}_{\mathrm{env}} \, \mathrm{d}a, \tag{366}
$$

and the energy transfer rate from the body to the surface internal power

$$
-\sum_{\alpha=1}^{n} \int_{\overline{\mathcal{S}}_{\mathrm{env}}} \mu^{\alpha} \boldsymbol{\jmath}^{\alpha} \cdot \boldsymbol{n} \, \mathrm{d}a. \tag{367}
$$

With these additional terms, the surface power imbalance (364) becomes

$$
\begin{aligned}
\Lambda^{\flat}_{\text{env}}(\{\dot{\varphi}^\alpha, \dot{\varphi}^\alpha_{\text{env}}\}_{\alpha=1}^n) = &\\
&\sum_{\alpha=1}^{n} \Bigg\{ \int\limits_{\overline{\mathcal{S}}_{\text{env}}} (\mu^\alpha_{\text{env}} \jmath^\alpha_{\text{env}} + \mu^\alpha \boldsymbol{\jmath}^\alpha \cdot \boldsymbol{n})\, \mathrm{d}a \\
&+ \int\limits_{\overline{\mathcal{S}}_{\text{env}}} (\xi^\alpha_{\mathcal{S}\,\text{env}} \dot{\varphi}^\alpha_{\text{env}} + {}^2\xi^\alpha_{\mathcal{S}\,\text{env}} \partial_n \dot{\varphi}^\alpha_{\text{env}} + {}^3\xi^\alpha_{\mathcal{S}\,\text{env}} \partial^2_n \dot{\varphi}^\alpha_{\text{env}})\, \mathrm{d}a \\
&- \int\limits_{\overline{\mathcal{S}}_{\text{env}}} (\xi^\alpha_{\mathcal{S}} \dot{\varphi}^\alpha + {}^2\xi^\alpha_{\mathcal{S}} \partial_n \dot{\varphi}^\alpha + {}^3\xi^\alpha_{\mathcal{S}} \partial^2_n \dot{\varphi}^\alpha)\, \mathrm{d}a \\
&+ \int\limits_{\overline{\partial\mathcal{S}}_{\text{env}}} (\xi^\alpha_{\partial\mathcal{S}\,\text{env}} \dot{\varphi}^\alpha_{\text{env}} + {}^2_n\xi^\alpha_{\partial\mathcal{S}\,\text{env}} \partial_n \dot{\varphi}^\alpha_{\text{env}} + {}^2_\nu\xi^\alpha_{\partial\mathcal{S}\,\text{env}} \partial_\nu \dot{\varphi}^\alpha_{\text{env}})\, \mathrm{d}\sigma \\
&- \int\limits_{\overline{\partial\mathcal{S}}_{\text{env}}} (\xi^\alpha_{\partial\mathcal{S}} \dot{\varphi}^\alpha + {}^2_n\xi^\alpha_{\partial\mathcal{S}} \partial_n \dot{\varphi}^\alpha + {}^2_\nu\xi^\alpha_{\partial\mathcal{S}} \partial_\nu \dot{\varphi}^\alpha)\, \mathrm{d}\sigma \\
&+ \int\limits_{\overline{\mathcal{C}}_{\text{env}}} (\xi^\alpha_{\mathcal{C}\,\text{env}} \dot{\varphi}^\alpha_{\text{env}} + {}^2_n\xi^\alpha_{\mathcal{C}\,\text{env}} \partial_n \dot{\varphi}^\alpha_{\text{env}} + {}^2_\nu\xi^\alpha_{\mathcal{C}\,\text{env}} \partial_\nu \dot{\varphi}^\alpha_{\text{env}})\, \mathrm{d}\sigma \\
&- \int\limits_{\overline{\mathcal{C}}_{\text{env}}} (\xi^\alpha_{\mathcal{C}} \dot{\varphi}^\alpha + {}^2_n\xi^\alpha_{\mathcal{C}} \partial_n \dot{\varphi}^\alpha + {}^2_\nu\xi^\alpha_{\mathcal{C}} \partial_\nu \dot{\varphi}^\alpha)\, \mathrm{d}\sigma \\
&+ \xi^\alpha_{\partial^2\mathcal{S}\,\text{env}} \dot{\varphi}^\alpha_{\text{env}}|_{\partial^2\mathcal{S}_{\text{env}}} + \xi^\alpha_{\partial\mathcal{C}\,\text{env}} \dot{\varphi}^\alpha_{\text{env}}|_{\partial\mathcal{C}_{\text{env}}} + \xi^\alpha_{\mathcal{O}\,\text{env}} \dot{\varphi}^\alpha_{\text{env}}|_{\mathcal{O}_{\text{env}}} \\
&- \xi^\alpha_{\partial^2\mathcal{S}} \dot{\varphi}^\alpha|_{\partial^2\mathcal{S}_{\text{env}}} - \xi^\alpha_{\partial\mathcal{C}} \dot{\varphi}^\alpha|_{\partial\mathcal{C}_{\text{env}}} - \xi^\alpha_{\mathcal{O}} \dot{\varphi}^\alpha|_{\mathcal{O}_{\text{env}}} \Bigg\} \geq 0. \qquad (368)
\end{aligned}
$$

3 Uncoupled Conditions

Instead of creating sufficient and necessary conditions to satisfy the environmental surface power imbalances (364) and (365) (or (368), for the conserved case), we develop a less general yet simple and sufficient conditions by uncoupling the environmental surface power imbalances.

We here require that the inequality direction is satisfied independently on $\mathcal{S}_\tau$, $\partial\mathcal{S}_\tau$, $\mathcal{C}_\tau$, $\partial^2\mathcal{S}_\tau$, $\partial\mathcal{C}_\tau$, and $\mathcal{O}_\tau$. Thus, by localization, we arrive at

$$
\begin{cases}
\xi^{\alpha}_{\mathcal{S}\,\text{env}}(\dot\varphi^{\alpha}_{\text{env}} - \dot\varphi^{\alpha})|_{\mathcal{S}_{\text{env}}} \geq 0, & {}^2\xi^{\alpha}_{\mathcal{S}\,\text{env}}(\partial_n\dot\varphi^{\alpha}_{\text{env}} - \partial_n\dot\varphi^{\alpha})|_{\mathcal{S}_{\text{env}}} \geq 0, \\
{}^3\xi^{\alpha}_{\mathcal{S}\,\text{env}}(\partial^2_n\dot\varphi^{\alpha}_{\text{env}} - \partial^2_n\dot\varphi^{\alpha})|_{\mathcal{S}_{\text{env}}} \geq 0, & \xi^{\alpha}_{\partial\mathcal{S}\,\text{env}}(\dot\varphi^{\alpha}_{\text{env}} - \dot\varphi^{\alpha})|_{\partial\mathcal{S}_{\text{env}}} \geq 0 \\
{}^2_n\xi^{\alpha}_{\partial\mathcal{S}\,\text{env}}(\partial_n\dot\varphi^{\alpha}_{\text{env}} - \partial_n\dot\varphi^{\alpha})|_{\partial\mathcal{S}_{\text{env}}} \geq 0, & {}^2_\nu\xi^{\alpha}_{\partial\mathcal{S}\,\text{env}}(\partial_\nu\dot\varphi^{\alpha}_{\text{env}} - \partial_\nu\dot\varphi^{\alpha})|_{\partial\mathcal{S}_{\text{env}}} \geq 0, \\
\xi^{\alpha}_{\mathcal{C}\,\text{env}}(\dot\varphi^{\alpha}_{\text{env}} - \dot\varphi^{\alpha})|_{\mathcal{C}_{\text{env}}} \geq 0, & {}^2_n\xi^{\alpha}_{\mathcal{C}\,\text{env}}(\partial_n\dot\varphi^{\alpha}_{\text{env}} - \partial_n\dot\varphi^{\alpha})|_{\mathcal{C}_{\text{env}}} \geq 0 \\
{}^2_\nu\xi^{\alpha}_{\mathcal{C}\,\text{env}}(\partial_\nu\dot\varphi^{\alpha}_{\text{env}} - \partial_\nu\dot\varphi^{\alpha})|_{\mathcal{C}_{\text{env}}} \geq 0, & \xi^{\alpha}_{\partial^2\mathcal{S}\,\text{env}}(\dot\varphi^{\alpha}_{\text{env}} - \dot\varphi^{\alpha})|_{\partial^2\mathcal{S}_{\text{env}}} \geq 0, \\
\xi^{\alpha}_{\partial\mathcal{C}\,\text{env}}(\dot\varphi^{\alpha}_{\text{env}} - \dot\varphi^{\alpha})|_{\partial\mathcal{C}_{\text{env}}} \geq 0, & \xi^{\alpha}_{\mathcal{O}\,\text{env}}(\dot\varphi^{\alpha}_{\text{env}} - \dot\varphi^{\alpha})|_{\mathcal{O}_{\text{env}}} \geq 0,
\end{cases}
\tag{369}
$$

for $1 \leq \alpha \leq n$ and

$$
\begin{cases}
\boldsymbol{t}_{\mathcal{S}\,\text{env}} \cdot (\boldsymbol{v}_{\text{env}} - \boldsymbol{v})|_{\mathcal{S}_{\text{env}}}| \geq 0, & {}^2\boldsymbol{t}_{\mathcal{S}\,\text{env}} \cdot (\partial_n\boldsymbol{v}_{\text{env}} - \partial_n\boldsymbol{v})|_{\mathcal{S}_{\text{env}}} \geq 0, \\
{}^3\boldsymbol{t}_{\mathcal{S}\,\text{env}} \cdot (\partial^2_n\boldsymbol{v}_{\text{env}} - \partial^2_n\boldsymbol{v})|_{\mathcal{S}_{\text{env}}} \geq 0, & \boldsymbol{t}_{\partial\mathcal{S}\,\text{env}} \cdot (\boldsymbol{v}_{\text{env}} - \boldsymbol{v})|_{\partial\mathcal{S}_{\text{env}}} \geq 0 \\
{}^2_n\boldsymbol{t}_{\partial\mathcal{S}\,\text{env}} \cdot (\partial_n\boldsymbol{v}_{\text{env}} - \partial_n\boldsymbol{v})|_{\partial\mathcal{S}_{\text{env}}} \geq 0, & {}^2_\nu\boldsymbol{t}_{\partial\mathcal{S}\,\text{env}} \cdot (\partial_\nu\boldsymbol{v}_{\text{env}} - \partial_\nu\boldsymbol{v})|_{\partial\mathcal{S}_{\text{env}}} \geq 0, \\
\boldsymbol{t}_{\mathcal{C}\,\text{env}} \cdot (\boldsymbol{v}_{\text{env}} - \boldsymbol{v})|_{\mathcal{C}_{\text{env}}} \geq 0, & {}^2_n\boldsymbol{t}_{\mathcal{C}\,\text{env}} \cdot (\partial_n\boldsymbol{v}_{\text{env}} - \partial_n\boldsymbol{v})|_{\mathcal{C}_{\text{env}}} \geq 0 \\
{}^2_\nu\boldsymbol{t}_{\mathcal{C}\,\text{env}} \cdot (\partial_\nu\boldsymbol{v}_{\text{env}} - \partial_\nu\boldsymbol{v})|_{\mathcal{C}_{\text{env}}} \geq 0, & \boldsymbol{t}_{\partial^2\mathcal{S}\,\text{env}} \cdot (\boldsymbol{v}_{\text{env}} - \boldsymbol{v})|_{\partial^2\mathcal{S}_{\text{env}}} \geq 0, \\
\boldsymbol{t}_{\partial\mathcal{C}\,\text{env}} \cdot (\boldsymbol{v}_{\text{env}} - \boldsymbol{v})|_{\partial\mathcal{C}_{\text{env}}} \geq 0, & \boldsymbol{t}_{\mathcal{O}\,\text{env}} \cdot (\boldsymbol{v}_{\text{env}} - \boldsymbol{v})|_{\mathcal{O}_{\text{env}}} \geq 0.
\end{cases}
\tag{370}
$$

Next, consider the case where the phase fields represent the concentration of conserved species. Thus, we have one additional condition arising from the first term in (369), that is

$$
\jmath^{\alpha}_{\text{env}}(\mu^{\alpha}_{\text{env}} - \mu^{\alpha}) \geq 0, \tag{371}
$$

for $1 \leq \alpha \leq n$. Note that in expression (371), we have used the environmental surface species balance (355), that is, $\jmath^{\alpha}_{\text{env}} = -\boldsymbol{\jmath} \cdot \boldsymbol{n}$.

Chapter 8
Boundary Conditions

In this chapter, we derive thermodynamically consistent boundary conditions arising from the environmental balance and imbalance postulates on surfaces. We consider that there may exist three types of boundaries that compose $\partial\mathcal{B}$. Here, $\mathcal{S}_{\text{nat}}$, $\mathcal{S}_{\text{ess}}$, and $\mathcal{S}_{\text{mix}}$ are the boundaries where the natural, essential, and mixed boundary conditions are prescribed, respectively. Additionally, we assume that $\mathcal{S}_{\text{nat}} \cup \mathcal{S}_{\text{ess}} \cup \mathcal{S}_{\text{nat}} \equiv \partial\mathcal{B}$ where $\mathcal{S}_{\text{nat}} \cap \mathcal{S}_{\text{ess}} = \mathcal{S}_{\text{nat}} \cap \mathcal{S}_{\text{mix}} = \mathcal{S}_{\text{mix}} \cap \mathcal{S}_{\text{ess}} = \varnothing$. Lastly, the following treatment is independent of constitutive equations.

1 Natural Boundary Conditions

To determine the natural boundary conditions, we use the environmental surface balances of microforces (350), forces (351), microtorques (352), and torques (353) which, by localization, render on $\mathcal{S}_{\text{nat}}$:

$$\begin{aligned}
\xi^{\alpha}_{\mathcal{S}\,\text{env}} &= (\boldsymbol{\xi}^{\alpha} - \operatorname{div} \boldsymbol{\Xi}^{\alpha} + \operatorname{div}^2 \mathbb{X}^{\alpha}) \cdot \boldsymbol{n} - \operatorname{div}_{\mathcal{S}} (\mathbf{P}_{\!n}[(\boldsymbol{\Xi}^{\alpha} - \operatorname{div} \mathbb{X}^{\alpha})[\boldsymbol{n}]]) \\
&\quad + \operatorname{div}_{\mathcal{S}} (\mathbf{P}_{\!n}[\operatorname{div}_{\mathcal{S}} (\mathbb{X}^{\alpha}[\boldsymbol{n}]\mathbf{P}_{\!n})]) - \operatorname{div}_{\mathcal{S}} (\mathbb{X}^{\alpha}[\boldsymbol{n}][\boldsymbol{n}] \cdot \mathbf{K}^{\top}), \qquad (372)\\
\boldsymbol{t}_{\mathcal{S}\,\text{env}} &= (\mathbf{T} - \operatorname{div} \mathbb{T} + \operatorname{div}^2 \mathbf{\mathbb{T}})[\boldsymbol{n}] - \operatorname{div}_{\mathcal{S}} ((\mathbb{T} - \operatorname{div} \mathbf{\mathbb{T}})[\boldsymbol{n}]\mathbf{P}_{\!n}) \\
&\quad + \operatorname{div}_{\mathcal{S}} (\operatorname{div}_{\mathcal{S}} (\mathbf{\mathbb{T}}[\boldsymbol{n}]\mathbf{P}_{\!n})\mathbf{P}_{\!n}) - \operatorname{div}_{\mathcal{S}} (\mathbf{\mathbb{T}}[\boldsymbol{n}][\boldsymbol{n}]\mathbf{K}^{\top}), \qquad (373)\\
\boldsymbol{\varpi}^{\alpha}_{\mathcal{S}\,\text{env}} &= (\boldsymbol{n} \otimes \boldsymbol{n})[(\boldsymbol{\Xi}^{\alpha} - \operatorname{div} \mathbb{X}^{\alpha})[\boldsymbol{n}] - \operatorname{div}_{\mathcal{S}} (\mathbb{X}^{\alpha}[\boldsymbol{n}]\mathbf{P}_{\!n})] \\
&\quad - \mathbb{X}^{\alpha}[\boldsymbol{n}][\boldsymbol{n}] \cdot \mathbf{K}^{\top}, \qquad (374)\\
\boldsymbol{m}_{\mathcal{S}\,\text{env}} &= \boldsymbol{n} \times ((\mathbb{T} - \operatorname{div} \mathbf{\mathbb{T}})[\boldsymbol{n}][\boldsymbol{n}] - \operatorname{div}_{\mathcal{S}} (\mathbf{\mathbb{T}}[\boldsymbol{n}]\mathbf{P}_{\!n})[\boldsymbol{n}]) \\
&\quad - 2 \operatorname{ax} (\operatorname{skw} (\mathbf{\mathbb{T}}[\boldsymbol{n}][\boldsymbol{n}]\mathbf{K}^{\top})^{\top}), \qquad (375)\\
\xi^{\alpha}_{\partial\mathcal{S}\,\text{env}} &= \boldsymbol{\nu} \cdot ((\boldsymbol{\Xi}^{\alpha} - \operatorname{div} \mathbb{X}^{\alpha})[\boldsymbol{n}] - \operatorname{div}_{\mathcal{S}} (\mathbb{X}^{\alpha}[\boldsymbol{n}]\mathbf{P}_{\!n})
\end{aligned}$$

L. Espath, *Mechanics and Geometry of Enriched Continua*,
https://doi.org/10.1007/978-3-031-28934-7_8

$$+\mathbb{X}^\alpha[\boldsymbol{n}][\boldsymbol{n}]\cdot\mathbf{K}^\top)-\partial_\sigma(\boldsymbol{\sigma}\cdot\mathbb{X}^\alpha[\boldsymbol{n}][\boldsymbol{\nu}]), \tag{376}$$

$$\boldsymbol{t}_{\partial\mathcal{S}\text{env}}=(\mathbb{T}-\operatorname{div}\mathbf{T})[\boldsymbol{n}][\boldsymbol{\nu}]-\operatorname{div}_{\mathcal{S}}((\mathbf{T}[\boldsymbol{n}])\mathbf{P}_n)[\boldsymbol{\nu}]$$

$$+\mathbf{T}[\boldsymbol{n}][\boldsymbol{n}][\mathbf{K}^\top[\boldsymbol{\nu}]]-\partial_\sigma(\mathbf{T}[\boldsymbol{n}][\boldsymbol{\nu}][\boldsymbol{\sigma}]), \tag{377}$$

$$\varpi^\alpha_{\partial\mathcal{S}\text{env}}=\mathbf{P}_\sigma[\mathbb{X}^\alpha[\boldsymbol{n}][\boldsymbol{\nu}]], \tag{378}$$

$$\boldsymbol{m}_{\partial\mathcal{S}\text{env}}=2\operatorname{ax}(\operatorname{skw}(\mathbf{T}[\boldsymbol{n}][\boldsymbol{\nu}]^\top))-\boldsymbol{\sigma}\times\mathbf{T}[\boldsymbol{n}][\boldsymbol{\nu}][\boldsymbol{\sigma}], \tag{379}$$

$$\xi^\alpha_{\mathcal{C}\,\text{env}}=\{\!\!\{\boldsymbol{\nu}\cdot((\Xi^\alpha-\operatorname{div}\mathbb{X}^\alpha)[\boldsymbol{n}]-\operatorname{div}_{\mathcal{S}}(\mathbb{X}^\alpha[\boldsymbol{n}]\mathbf{P}_n)$$

$$+\mathbb{X}^\alpha[\boldsymbol{n}][\boldsymbol{n}]\cdot\mathbf{K}^\top)-\partial_\sigma(\boldsymbol{\sigma}\cdot\mathbb{X}^\alpha[\boldsymbol{n}][\boldsymbol{\nu}])\}\!\!\}, \tag{380}$$

$$\boldsymbol{t}_{\mathcal{C}\text{env}}=\{\!\!\{(\mathbb{T}-\operatorname{div}\mathbf{T})[\boldsymbol{n}][\boldsymbol{\nu}]-\operatorname{div}_{\mathcal{S}}(\mathbf{T}[\boldsymbol{n}]\mathbf{P}_n)[\boldsymbol{\nu}]$$

$$+\mathbf{T}[\boldsymbol{n}][\boldsymbol{n}][\mathbf{K}^\top[\boldsymbol{\nu}]]-\partial_\sigma(\mathbf{T}[\boldsymbol{n}][\boldsymbol{\nu}][\boldsymbol{\sigma}])\}\!\!\}, \tag{381}$$

$$\varpi^\alpha_{\mathcal{C}\,\text{env}}=\{\!\!\{\mathbf{P}_\sigma[\mathbb{X}^\alpha[\boldsymbol{n}][\boldsymbol{\nu}]]\}\!\!\}, \tag{382}$$

$$\boldsymbol{m}_{\mathcal{C}\text{env}}=2\operatorname{ax}(\operatorname{skw}(\{\!\!\{\mathbf{T}[\boldsymbol{n}][\boldsymbol{\nu}]\}\!\!\})^\top)-\boldsymbol{\sigma}\times\{\!\!\{\mathbf{T}[\boldsymbol{n}][\boldsymbol{\nu}][\boldsymbol{\sigma}]\}\!\!\}, \tag{383}$$

$$\xi^\alpha_{\partial^2\mathcal{S}\text{env}}=[\![\boldsymbol{\sigma}\cdot\mathbb{X}^\alpha[\boldsymbol{n}][\boldsymbol{\nu}]]\!], \tag{384}$$

$$\boldsymbol{t}_{\partial^2\mathcal{S}\text{env}}=[\![\mathbf{T}[\boldsymbol{n}][\boldsymbol{\nu}][\boldsymbol{\sigma}]]\!], \tag{385}$$

$$\xi^\alpha_{\partial\mathcal{C}\text{env}}=]\!]\{\!\!\{\boldsymbol{\sigma}\cdot\mathbb{X}^\alpha[\boldsymbol{n}][\boldsymbol{\nu}]\}\!\!\}[\![, \tag{386}$$

$$\boldsymbol{t}_{\partial\mathcal{C}\text{env}}=]\!]\{\!\!\{\mathbf{T}[\boldsymbol{n}][\boldsymbol{\nu}][\boldsymbol{\sigma}]\}\!\!\}[\![, \tag{387}$$

$$\xi^\alpha_{\mathcal{O}\text{env}}=[\![\{\!\!\{\boldsymbol{\sigma}\cdot\mathbb{X}^\alpha[\boldsymbol{n}][\boldsymbol{\nu}]\}\!\!\}]\!], \tag{388}$$

$$\boldsymbol{t}_{\mathcal{O}\text{env}}=[\![\{\!\!\{\mathbf{T}[\boldsymbol{n}][\boldsymbol{\nu}][\boldsymbol{\sigma}]\}\!\!\}]\!], \tag{389}$$

for $1\le\alpha\le n$. Additionally, in considering the case where the phase fields represent the concentration of conserved species, from the environmental surface species balance (354), we have that $\jmath^\alpha_{\text{env}}=-\boldsymbol{\jmath}\cdot\boldsymbol{n}$ on $\mathcal{S}_{\text{nat}}$.

2 Essential Boundary Conditions

Next, we define the essential boundary conditions as the trivial nondissipative solution of (369) and (370), respectively, that is, on $\mathcal{S}_{\text{ess}}$:

$$(\dot\varphi^\alpha_{\text{env}}-\dot\varphi^\alpha)|_{\mathcal{S}_{\text{env}}}=0,\qquad(\boldsymbol{v}_{\text{env}}-\boldsymbol{v})|_{\mathcal{S}_{\text{env}}}=\mathbf{0}, \tag{390}$$

$$(\partial_n\dot\varphi^\alpha_{\text{env}}-\partial_n\dot\varphi^\alpha)|_{\mathcal{S}_{\text{env}}}=0,\qquad(\partial_n\boldsymbol{v}_{\text{env}}-\partial_n\boldsymbol{v})|_{\mathcal{S}_{\text{env}}}=\mathbf{0}, \tag{391}$$

$$(\partial^2_n\dot\varphi^\alpha_{\text{env}}-\partial^2_n\dot\varphi^\alpha)|_{\mathcal{S}_{\text{env}}}=0,\qquad(\partial^2_n\boldsymbol{v}_{\text{env}}-\partial^2_n\boldsymbol{v})|_{\mathcal{S}_{\text{env}}}=\mathbf{0}, \tag{392}$$

$$(\dot\varphi^\alpha_{\text{env}}-\dot\varphi^\alpha)|_{\partial\mathcal{S}_{\text{env}}}=0\qquad(\boldsymbol{v}_{\text{env}}-\boldsymbol{v})|_{\partial\mathcal{S}_{\text{env}}}=\mathbf{0}, \tag{393}$$

$$(\partial_n\dot\varphi^\alpha_{\text{env}}-\partial_n\dot\varphi^\alpha)|_{\partial\mathcal{S}_{\text{env}}}=0,\qquad(\partial_n\boldsymbol{v}_{\text{env}}-\partial_n\boldsymbol{v})|_{\partial\mathcal{S}_{\text{env}}}=\mathbf{0}, \tag{394}$$

$$(\partial_\nu\dot\varphi^\alpha_{\text{env}}-\partial_\nu\dot\varphi^\alpha)|_{\partial\mathcal{S}_{\text{env}}}=0,\qquad(\partial_\nu\boldsymbol{v}_{\text{env}}-\partial_\nu\boldsymbol{v})|_{\partial\mathcal{S}_{\text{env}}}=\mathbf{0}, \tag{395}$$

$$(\dot\varphi^\alpha_{\text{env}}-\dot\varphi^\alpha)|_{\mathcal{C}_{\text{env}}}=0,\qquad(\boldsymbol{v}_{\text{env}}-\boldsymbol{v})|_{\mathcal{C}_{\text{env}}}=\mathbf{0}, \tag{396}$$

$$(\partial_n\dot\varphi^\alpha_{\text{env}}-\partial_n\dot\varphi^\alpha)|_{\mathcal{C}_{\text{env}}}=0\qquad(\partial_n\boldsymbol{v}_{\text{env}}-\partial_n\boldsymbol{v})|_{\mathcal{C}_{\text{env}}}=\mathbf{0}, \tag{397}$$

$$(\partial_\nu \dot{\varphi}^\alpha_{\text{env}} - \partial_\nu \dot{\varphi}^\alpha)|_{\mathcal{C}_{\text{env}}} = 0, \qquad (\partial_\nu \boldsymbol{v}_{\text{env}} - \partial_\nu \boldsymbol{v})|_{\mathcal{C}_{\text{env}}} = \mathbf{0}, \tag{398}$$

$$(\dot{\varphi}^\alpha_{\text{env}} - \dot{\varphi}^\alpha)|_{\partial^2 \mathcal{S}_{\text{env}}} = 0, \qquad (\boldsymbol{v}_{\text{env}} - \boldsymbol{v})|_{\partial^2 \mathcal{S}_{\text{env}}} = \mathbf{0}, \tag{399}$$

$$(\dot{\varphi}^\alpha_{\text{env}} - \dot{\varphi}^\alpha)|_{\partial \mathcal{C}_{\text{env}}} = 0, \qquad (\boldsymbol{v}_{\text{env}} - \boldsymbol{v})|_{\partial \mathcal{C}_{\text{env}}} = \mathbf{0}, \tag{400}$$

$$(\dot{\varphi}^\alpha_{\text{env}} - \dot{\varphi}^\alpha)|_{\mathcal{O}_{\text{env}}} = 0, \qquad (\boldsymbol{v}_{\text{env}} - \boldsymbol{v})|_{\mathcal{O}_{\text{env}}} = \mathbf{0}, \tag{401}$$

for $1 \le \alpha \le n$.

3 Mixed Boundary Conditions

Finally, mixed boundary conditions also arise from (369) and (370). We opt for the simplest linear solution for the environmental component of these expressions, rendering on $\mathcal{S}_{\text{mix}}$:

$$\xi^\alpha_{\mathcal{S}\,\text{env}} = a_1(\dot{\varphi}^\alpha_{\text{env}} - \dot{\varphi}^\alpha)|_{\mathcal{S}_{\text{env}}}, \qquad \boldsymbol{t}_{\mathcal{S}\text{env}} = b_1(\boldsymbol{v}_{\text{env}} - \boldsymbol{v})|_{\mathcal{S}_{\text{env}}}, \tag{402}$$

$${}^2\xi^\alpha_{\mathcal{S}\,\text{env}} = a_2(\partial_n \dot{\varphi}^\alpha_{\text{env}} - \partial_n \dot{\varphi}^\alpha)|_{\mathcal{S}_{\text{env}}}, \qquad {}^2\boldsymbol{t}_{\mathcal{S}\text{env}} = b_2(\partial_n \boldsymbol{v}_{\text{env}} - \partial_n \boldsymbol{v})|_{\mathcal{S}_{\text{env}}}, \tag{403}$$

$${}^3\xi^\alpha_{\mathcal{S}\,\text{env}} = a_3(\partial_n^2 \dot{\varphi}^\alpha_{\text{env}} - \partial_n^2 \dot{\varphi}^\alpha)|_{\mathcal{S}_{\text{env}}}, \qquad {}^3\boldsymbol{t}_{\mathcal{S}\text{env}} = b_3(\partial_n^2 \boldsymbol{v}_{\text{env}} - \partial_n^2 \boldsymbol{v})|_{\mathcal{S}_{\text{env}}}, \tag{404}$$

$$\xi^\alpha_{\partial\mathcal{S}\,\text{env}} = a_4(\dot{\varphi}^\alpha_{\text{env}} - \dot{\varphi}^\alpha)|_{\partial\mathcal{S}_{\text{env}}}, \qquad \boldsymbol{t}_{\partial\mathcal{S}\text{env}} = b_4(\boldsymbol{v}_{\text{env}} - \boldsymbol{v})|_{\partial\mathcal{S}_{\text{env}}}, \tag{405}$$

$${}^2_n\xi^\alpha_{\partial\mathcal{S}\,\text{env}} = a_5(\partial_n \dot{\varphi}^\alpha_{\text{env}} - \partial_n \dot{\varphi}^\alpha)|_{\partial\mathcal{S}_{\text{env}}}, \qquad {}^2_n\boldsymbol{t}_{\partial\mathcal{S}\text{env}} = b_5(\partial_n \boldsymbol{v}_{\text{env}} - \partial_n \boldsymbol{v})|_{\partial\mathcal{S}_{\text{env}}}, \tag{406}$$

$${}^2_\nu\xi^\alpha_{\partial\mathcal{S}\,\text{env}} = a_6(\partial_\nu \dot{\varphi}^\alpha_{\text{env}} - \partial_\nu \dot{\varphi}^\alpha)|_{\partial\mathcal{S}_{\text{env}}}, \qquad {}^2_\nu\boldsymbol{t}_{\partial\mathcal{S}\text{env}} = b_6(\partial_\nu \boldsymbol{v}_{\text{env}} - \partial_\nu \boldsymbol{v})|_{\partial\mathcal{S}_{\text{env}}}, \tag{407}$$

$$\xi^\alpha_{\mathcal{C}\,\text{env}} = a_7(\dot{\varphi}^\alpha_{\text{env}} - \dot{\varphi}^\alpha)|_{\mathcal{C}_{\text{env}}}, \qquad \boldsymbol{t}_{\mathcal{C}\text{env}} = b_7(\boldsymbol{v}_{\text{env}} - \boldsymbol{v})|_{\mathcal{C}_{\text{env}}}, \tag{408}$$

$${}^2_n\xi^\alpha_{\mathcal{C}\,\text{env}} = a_8(\partial_n \dot{\varphi}^\alpha_{\text{env}} - \partial_n \dot{\varphi}^\alpha)|_{\mathcal{C}_{\text{env}}}, \qquad {}^2_n\boldsymbol{t}_{\mathcal{C}\text{env}} = b_8(\partial_n \boldsymbol{v}_{\text{env}} - \partial_n \boldsymbol{v})|_{\mathcal{C}_{\text{env}}}, \tag{409}$$

$${}^2_\nu\xi^\alpha_{\mathcal{C}\,\text{env}} = a_9(\partial_\nu \dot{\varphi}^\alpha_{\text{env}} - \partial_\nu \dot{\varphi}^\alpha)|_{\mathcal{C}_{\text{env}}}, \qquad {}^2_\nu\boldsymbol{t}_{\mathcal{C}\text{env}} = b_9(\partial_\nu \boldsymbol{v}_{\text{env}} - \partial_\nu \boldsymbol{v})|_{\mathcal{C}_{\text{env}}}, \tag{410}$$

$$\xi^\alpha_{\partial^2\mathcal{S}\,\text{env}} = a_{10}(\dot{\varphi}^\alpha_{\text{env}} - \dot{\varphi}^\alpha)|_{\partial^2\mathcal{S}_{\text{env}}}, \qquad \boldsymbol{t}_{\partial^2\mathcal{S}\text{env}} = b_{10}(\boldsymbol{v}_{\text{env}} - \boldsymbol{v})|_{\partial^2\mathcal{S}_{\text{env}}}, \tag{411}$$

$$\xi^\alpha_{\partial\mathcal{C}\,\text{env}} = a_{11}(\dot{\varphi}^\alpha_{\text{env}} - \dot{\varphi}^\alpha)|_{\partial\mathcal{C}_{\text{env}}}, \qquad \boldsymbol{t}_{\partial\mathcal{C}\text{env}} = b_{11}(\boldsymbol{v}_{\text{env}} - \boldsymbol{v})|_{\partial\mathcal{C}_{\text{env}}}, \tag{412}$$

$$\xi^\alpha_{\mathcal{O}\,\text{env}} = a_{12}(\dot{\varphi}^\alpha_{\text{env}} - \dot{\varphi}^\alpha)|_{\mathcal{O}_{\text{env}}}, \qquad \boldsymbol{t}_{\mathcal{O}\text{env}} = b_{12}(\boldsymbol{v}_{\text{env}} - \boldsymbol{v})|_{\mathcal{O}_{\text{env}}}, \tag{413}$$

where $a_i > 0$ and $b_i > 0$ for $1 \le i \le 12$ and $1 \le \alpha \le n$.

Part V
A Special Theory

Chapter 9
Bulk-Surface Dynamics

In this chapter, based on the work by Espath [1], we aim at detailing the underlying rational mechanics of dynamic boundary conditions proposed by Fischer, Maass, & Dieterich [62], Goldstein, Miranville, & Schimperna [63], and Knopf, Lam, Liu & Metzger, [64]. As a byproduct, we generalize these theories. These types of dynamic boundary conditions are described by the coupling between the bulk and surface partial differential equations for phase fields. Our point of departure within this continuum framework is the principle of virtual powers postulated on an arbitrary part $\mathcal{P}$ where the boundary $\partial\mathcal{P}$ may lose smoothness. That is, the normal field may be discontinuous along an edge $\partial^2\mathcal{P}$. However, the edges characterizing the discontinuity of the normal field are considered smooth. Our results may be summarized as follows. We provide a generalized version of the principle of virtual powers for the bulk-surface coupling along with a generalized version of the partwise free-energy imbalance. Next, we derive the explicit form of the surface and edge microtractions along with the field equations for the bulk and surface phase fields. The final set of field equations somewhat resembles the Cahn–Hilliard equation for both the bulk and surface. Moreover, we provide a suitable set of constitutive relations and thermodynamically consistent boundary conditions. In [64], a mixed type of boundary condition for the chemical potentials is proposed for the model in [62, 63]. In addition to this boundary condition, we also include this type of mixed boundary condition for the microstructure, that is, the phase fields. Lastly, we derive the Lyapunov decay relations for these mixed types of boundary conditions for both the microstructure and chemical potential.

L. Espath, *Mechanics and Geometry of Enriched Continua*,
https://doi.org/10.1007/978-3-031-28934-7_9

1 Introduction

1.1 Synopsis of Purely Variational Models

The model proposed in [62–64] on a body $\mathcal{P}$ with boundary $\partial\mathcal{P}$ for the underlying free-energy functional

$$\begin{aligned}\Psi[\varphi_{\mathcal{P}},\varphi_{\partial\mathcal{P}}] &= \int_{\mathcal{P}} \psi_{\mathcal{P}}\,\mathrm{d}v + \int_{\partial\mathcal{P}} \psi_{\partial\mathcal{P}}\,\mathrm{d}a,\\ &= \int_{\mathcal{P}} \Big(\tfrac{1}{\epsilon} f(\varphi_{\mathcal{P}}) + \tfrac{\epsilon}{2}|\mathrm{grad}\,\varphi_{\mathcal{P}}|^2\Big)\,\mathrm{d}v + \int_{\partial\mathcal{P}} \Big(\tfrac{1}{\delta} g(\varphi_{\partial\mathcal{P}}) + \tfrac{\iota\delta}{2}|\mathrm{grad}_{\mathcal{S}}\varphi_{\partial\mathcal{P}}|^2\Big)\,\mathrm{d}a,\end{aligned} \tag{414}$$

reads

$$\begin{cases} \dot{\varphi}_{\mathcal{P}} = m_{\mathcal{P}}\Delta\mu_{\mathcal{P}}, & \text{in } \mathcal{P},\\ \mu_{\mathcal{P}} = -\epsilon\Delta\varphi_{\mathcal{P}} + \frac{1}{\epsilon} f'(\varphi_{\mathcal{P}}), & \text{in } \mathcal{P},\\ \dot{\varphi}_{\partial\mathcal{P}} = m_{\partial\mathcal{P}}\Delta_{\mathcal{S}}\mu_{\partial\mathcal{P}} - \beta m_{\mathcal{P}}\partial_n\mu_{\mathcal{P}}, & \text{on } \partial\mathcal{P},\\ \mu_{\partial\mathcal{P}} = -\iota\delta\Delta_{\mathcal{S}}\varphi_{\partial\mathcal{P}} + \frac{1}{\delta} g'(\varphi_{\partial\mathcal{P}}) + \epsilon\partial_n\varphi_{\mathcal{P}}, & \text{on } \partial\mathcal{P},\\ \varphi_{\mathcal{P}} = \varphi_{\partial\mathcal{P}}, & \text{on } \partial\mathcal{P},\\ \partial_n\mu_{\mathcal{P}} = \frac{1}{L}(\beta\mu_{\partial\mathcal{P}} - \mu_{\mathcal{P}}), & \text{on } \partial\mathcal{P}. \end{cases} \tag{415}$$

Here, $\psi_{\mathcal{P}}$ and $\psi_{\partial\mathcal{P}}$ represent the bulk and surface free-energy densities, respectively. Δ and $\Delta_{\mathcal{S}}$ are the Laplace and Laplace–Beltrami operators, respectively. The Laplace–Beltrami operator may be written as

$$\Delta_{\mathcal{S}}\kappa := \mathrm{div}_{\mathcal{S}}\mathrm{grad}_{\mathcal{S}}\kappa = \mathrm{grad}\,(\mathbf{P}_{n}[\mathrm{grad}\,\kappa]) : \mathbf{P}_{n}. \tag{416}$$

Also, $\varphi_{\mathcal{P}}$ and $\varphi_{\partial\mathcal{P}}$ are the bulk and surface conserved phase fields, $\mu_{\mathcal{P}}$ and $\mu_{\partial\mathcal{P}}$ are the bulk and surface chemical potentials, f and g are the bulk and surface potentials, and $m_{\mathcal{P}}$ and $m_{\partial\mathcal{P}}$ are the bulk and surface mobility coefficients. Lastly, ϵ, δ, ι, β and L are real positive constant parameters. It is important to what follows to note that (415) has the Cahn–Hilliard type of structure for both the bulk and surface.

1.2 Synopsis of this Work

In continuum mechanics, it is customary to isolate an arbitrary part $\mathcal{P}$ from a body $\mathcal{B}$ to describe the interactions between $\mathcal{P}$ and adjacent parts of $\mathcal{B}$ to establish balance laws. That is to say, the action of $\mathcal{B}\setminus\mathcal{P}$ on $\mathcal{P}$ is represented through surface tractions and normal fluxes. This is probably the most used concept in structural mechanics. We here abandon this hypothesis and consider that interactions between $\mathcal{P}$ and adjacent parts of $\mathcal{B}$ are described by additional evolution equations on $\partial\mathcal{P}$. This ultimately implies that the boundary conditions on $\mathcal{B}$ are defined through a partial differential equation on $\partial\mathcal{B}$. We may however limit the dynamic response of the environment to a certain region of $\partial\mathcal{B}$ instead of considering that the entire surrounding environment is dynamic. Note that since balances do not depend on material idealizations, we separate balance equations from constitutive response functions.

In Fig. 1, $\mathcal{B}$ denotes a region of a three-dimensional point space $\mathcal{E}$ where $\mathcal{P} \subseteq \mathcal{B}$ is an arbitrary subregion of $\mathcal{B}$ with a closed surface boundary $\partial\mathcal{P}$ oriented by an outward unit normal $\boldsymbol{n}$ at $\boldsymbol{x} \in \partial\mathcal{P}$. The surface $\partial\mathcal{P}$ may lose smoothness along a curve, namely an edge $\partial^2\mathcal{P}$. In a neighborhood of an edge $\partial^2\mathcal{P}$, two smooth surfaces $\partial\mathcal{P}^{\pm}$ are defined. The limiting unit normals of $\partial\mathcal{P}^{\pm}$ at $\partial^2\mathcal{P}$ are denoted by the pair $\{\boldsymbol{n}^+, \boldsymbol{n}^-\}$. The pair of unit normals characterizes the edge $\partial^2\mathcal{P}$. Similarly, the limiting outward unit tangent-normal of $\partial\mathcal{P}^{\pm}$ at $\partial^2\mathcal{P}$ are $\{\boldsymbol{\nu}^+, \boldsymbol{\nu}^-\}$. Additionally, $\partial^2\mathcal{P}$ is oriented by the unit tangent $\boldsymbol{\sigma} := \boldsymbol{\sigma}^+$ such that $\boldsymbol{\sigma}^+ := \boldsymbol{n}^+ \times \boldsymbol{\nu}^+$.

In this work, we propose a continuum theory with two kinematical processes, a bulk $\varphi_{\mathcal{P}}$ and a surface $\varphi_{\partial\mathcal{P}}$ fields on $\mathcal{P}$ and $\partial\mathcal{P}$, respectively. Within this framework, each of

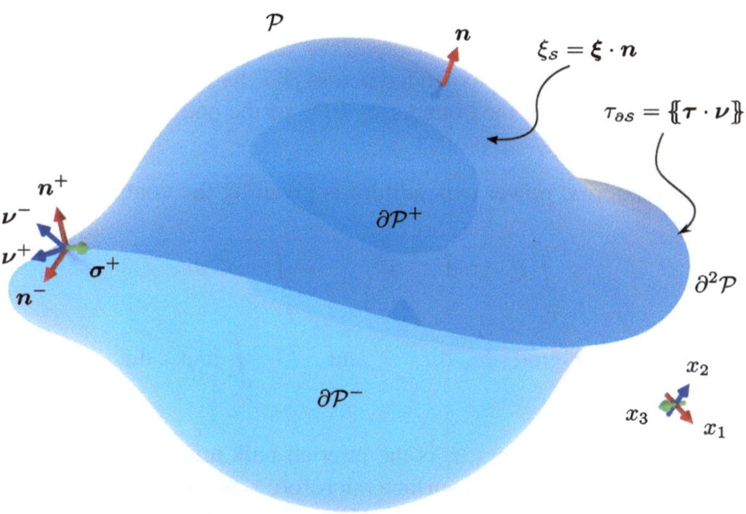

Fig. 1 (Computer Aided Design drawing by Luis Espath and renderization by Gabriel Nogueira de Castro, from [1], licensed under CC-BY 4.0). Part $\mathcal{P}$ with nonsmooth boundary surface $\partial\mathcal{P}^{\pm}$ oriented by the unit normal $\boldsymbol{n}$ with the outward unit tangent-normal $\boldsymbol{\nu}^{\pm}$ at the smooth boundary-edge $\partial^2\mathcal{P}$ oriented by the unit tangent $\boldsymbol{\sigma} := \boldsymbol{n} \times \boldsymbol{\nu}$. The surface $\partial\mathcal{P}$ lacks smoothness at an edge $\partial^2\mathcal{P}$

$$\int_{\mathcal{P}} \gamma\dot{\varphi}_{\mathcal{P}}\,\mathrm{d}v, \qquad \int_{\partial\mathcal{P}} \xi_{\mathcal{S}}\dot{\varphi}_{\mathcal{P}}\,\mathrm{d}a,$$
$$\int_{\partial\mathcal{P}} \zeta\dot{\varphi}_{\partial\mathcal{P}}\,\mathrm{d}a, \qquad -\int_{\partial\mathcal{P}} \xi_{\mathcal{S}}\dot{\varphi}_{\partial\mathcal{P}}\,\mathrm{d}a, \qquad \text{and} \qquad \int_{\partial^2\mathcal{P}} \tau_{\partial\mathcal{S}}\dot{\varphi}_{\partial\mathcal{P}}\,\mathrm{d}\sigma \tag{417}$$

represents an external form of power expenditure, where γ is the external bulk microforce, $\xi_{\mathcal{S}}$ is the surface microtraction, ζ is the external surface microforce, and $\tau_{\partial\mathcal{S}}$ is the edge microtraction. These power expenditures may be described as follows.

- $\gamma\dot{\varphi}_{\mathcal{P}}$ represents the power expended on the atoms of $\mathcal{P}$ by sources external to the body $\mathcal{P}$;
- $\xi_{\mathcal{S}}\dot{\varphi}_{\mathcal{P}}$ represents the power expended across $\partial\mathcal{P}$ by configurations neighboring the boundary of the body $\partial\mathcal{P}$ and exterior to $\mathcal{P}$;
- $\zeta\dot{\varphi}_{\partial\mathcal{P}}$ represents the power expended on the atoms of $\partial\mathcal{P}$ by sources external to the boundary of the body $\partial\mathcal{P}$ and not originated from $\mathcal{P}$;
- $-\xi_{\mathcal{S}}\dot{\varphi}_{\partial\mathcal{P}}$ represents the power expended on the atoms of $\partial\mathcal{P}$ by sources external to the boundary of the body and originated from $\mathcal{P}$;
- $\tau_{\partial\mathcal{S}}\dot{\varphi}_{\partial\mathcal{P}}$ represents the power expended across $\partial^2\mathcal{P}$ by configurations neighboring the common boundary of the boundaries $\partial\mathcal{P}^{\pm}$ of the body $\mathcal{P}$ and exterior to both $\partial\mathcal{P}$ and $\mathcal{P}$.

Conversely, the internal power expenditure is given by the contribution of the following terms

$$\int_{\mathcal{P}} \boldsymbol{\xi}\cdot\operatorname{grad}\dot{\varphi}_{\mathcal{P}}\,\mathrm{d}v, \qquad -\int_{\mathcal{P}} \pi\dot{\varphi}_{\mathcal{P}}\,\mathrm{d}v,$$
$$\int_{\partial\mathcal{P}} \boldsymbol{\tau}\cdot\operatorname{grad}_{\mathcal{S}}\dot{\varphi}_{\partial\mathcal{P}}\,\mathrm{d}a, \qquad \text{and} \qquad -\int_{\partial\mathcal{P}} \varpi\dot{\varphi}_{\partial\mathcal{P}}\,\mathrm{d}a, \tag{418}$$

where $\boldsymbol{\xi}$ is the bulk microstress, π is the internal bulk microforce, $\boldsymbol{\tau}$ is the surface microstress, and ϖ is the internal surface microforce. We then base our treatment on the virtual power principle formulation by Gurtin [42] and Fried & Gurtin [20]. These works represent our point of departure to propose a generalized bulk-surface version of this principle. Through this suitable principle of virtual powers, we arrive at the microtractions presented in the external power, $\xi_{\mathcal{S}}$ and $\tau_{\partial\mathcal{S}}$, and the field equations.

Next, given the bulk and surface species fluxes, $\boldsymbol{\jmath}_{\mathcal{P}}$ and $\boldsymbol{\jmath}_{\partial\mathcal{P}}$, and the bulk and surface external rates of species production, $s_{\mathcal{P}}$ and $s_{\partial\mathcal{P}}$, we postulate the partwise species balances for $\mathcal{P}$ and $\partial\mathcal{P}$, where the balance on $\partial\mathcal{P}$ is supplemented by a contribution originated from $\mathcal{P}$ and given by

$$\int_{\partial\mathcal{P}} \beta\, \jmath_{\mathcal{P}} \cdot \boldsymbol{n}\, \mathrm{d}a. \tag{419}$$

Then, with a suitable free-energy imbalance, we account for the rate at which energy is transferred to $\mathcal{P}$ and $\partial\mathcal{P}$ due to species transport to determine the constitutive relations and arrive at the following set of equations

$$\begin{cases} \dot{\varphi}_{\mathcal{P}} = s_{\mathcal{P}} - \mathrm{div}\, \jmath_{\mathcal{P}}, & \text{in } \mathcal{P}, \\ \mu_{\mathcal{P}} = -\mathrm{div}\boldsymbol{\xi} - \gamma + \partial_{\varphi_{\mathcal{P}}} \psi_{\mathcal{P}}, & \text{in } \mathcal{P}, \\ \dot{\varphi}_{\partial\mathcal{P}} = \beta\, \jmath_{\mathcal{P}} \cdot \boldsymbol{n} + s_{\partial\mathcal{P}} - \mathrm{div}_{\mathcal{S}} \jmath_{\partial\mathcal{P}} - 2K \jmath_{\partial\mathcal{P}} \cdot \boldsymbol{n}, & \text{on } \partial\mathcal{P}, \\ \mu_{\partial\mathcal{P}} = -\mathrm{div}_{\mathcal{S}} \boldsymbol{\tau} - 2K \boldsymbol{\tau} \cdot \boldsymbol{n} - \zeta + \boldsymbol{\xi} \cdot \boldsymbol{n} + \partial_{\varphi_{\partial\mathcal{P}}} \psi_{\partial\mathcal{P}}, & \text{on } \partial\mathcal{P}, \end{cases} \tag{420}$$

and

$$\begin{cases} \boldsymbol{\xi} = \partial_{\mathrm{grad}\,\varphi_{\mathcal{P}}} \psi_{\mathcal{P}}, & \text{in } \mathcal{P}, \\ \jmath_{\mathcal{P}} = -\boldsymbol{M}_{\mathcal{P}}[\mathrm{grad}\, \mu_{\mathcal{P}}], & \text{in } \mathcal{P}, \\ \boldsymbol{\tau} = \partial_{\mathrm{grad}_{\mathcal{S}}\varphi_{\partial\mathcal{P}}} \psi_{\partial\mathcal{P}}, & \text{on } \partial\mathcal{P}, \\ \jmath_{\partial\mathcal{P}} = -\boldsymbol{M}_{\partial\mathcal{P}}[\mathrm{grad}_{\mathcal{S}} \mu_{\partial\mathcal{P}}], & \text{on } \partial\mathcal{P}, \end{cases} \tag{421}$$

where $\boldsymbol{M}_{\mathcal{P}}$ and $\boldsymbol{M}_{\partial\mathcal{P}}$ are the bulk and surface mobility tensors, respectively.

Also, note that the appearance of bulk microstress $\boldsymbol{\xi}$ in the surface chemical potential $\mu_{\partial\mathcal{P}}$, in expression $(420)_4$, results from the coupling at the principle of virtual powers' level, whereas the presence of the bulk species flux $\jmath_{\mathcal{P}}$ in the surface species time derivative $\dot{\varphi}_{\partial\mathcal{P}}$, in expression $(420)_3$, results from the coupling at the surface species balance's level.

Aside from the fact that we present a new version of the principle of virtual powers and free-energy imbalance, there are three key differences between our continuum framework and previous works on dynamic boundary conditions. First, our theory is based on underlying mechanical principles. Second, our theory generalizes the resulting equations in [63, 64]. Third, we consider that the boundary $\partial\mathcal{P}$ may be endowed with a discontinuous normal field, allowing the assignment of edge microtractions. Lastly, in [64], a mixed type of boundary condition for the chemical potentials is proposed for the model in [62, 63]. In addition to this boundary condition, we also include this type of mixed boundary condition for the microstructures, that is, the phase fields.

2 Virtual Power Principle

We are now in a position to postulate the principle of virtual powers. Considering the power expenditures discussed in the previous section, the principle reads

$$\mathcal{V}_{\mathrm{ext}}(\mathcal{P}, \partial\mathcal{P}; \chi_{\mathcal{P}}, \chi_{\partial\mathcal{P}}) = \mathcal{V}_{\mathrm{int}}(\mathcal{P}, \partial\mathcal{P}; \chi_{\mathcal{P}}, \chi_{\partial\mathcal{P}}), \tag{422}$$

where $\chi_{\mathcal{P}}$ and $\chi_{\partial\mathcal{P}}$ are two sufficiently smooth virtual fields defined, respectively, on $\mathcal{P}$ and $\partial\mathcal{P}$. The external and internal virtual power are, respectively, given by

$$\begin{aligned}\mathcal{V}_{\text{ext}}(\mathcal{P},\partial\mathcal{P};\chi_{\mathcal{P}},\chi_{\partial\mathcal{P}}) &= \int\limits_{\mathcal{P}} \gamma\chi_{\mathcal{P}}\,\mathrm{d}v + \int\limits_{\partial\mathcal{P}} (\zeta - \xi_{\mathcal{S}})\chi_{\partial\mathcal{P}}\,\mathrm{d}a \\ &\quad + \int\limits_{\partial\mathcal{P}} \xi_{\mathcal{S}}\chi_{\mathcal{P}}\,\mathrm{d}a + \int\limits_{\partial^2\mathcal{P}} \tau_{\partial\mathcal{S}}\chi_{\partial\mathcal{P}}\,\mathrm{d}\sigma,\end{aligned} \tag{423}$$

and

$$\begin{aligned}\mathcal{V}_{\text{int}}(\mathcal{P},\partial\mathcal{P};\chi_{\mathcal{P}},\chi_{\partial\mathcal{P}}) &= \int\limits_{\mathcal{P}} \boldsymbol{\xi}\cdot\operatorname{grad}\chi_{\mathcal{P}}\,\mathrm{d}v - \int\limits_{\mathcal{P}} \pi\chi_{\mathcal{P}}\,\mathrm{d}v \\ &\quad + \int\limits_{\partial\mathcal{P}} \boldsymbol{\tau}\cdot\operatorname{grad}_{\mathcal{S}}\chi_{\partial\mathcal{P}}\,\mathrm{d}a - \int\limits_{\partial\mathcal{P}} \varpi\chi_{\partial\mathcal{P}}\,\mathrm{d}a.\end{aligned} \tag{424}$$

Next, we aim at deriving the explicit forms of the surface microtraction $\xi_{\mathcal{S}}$ and the edge microtraction $\tau_{\partial\mathcal{S}}$. Noting that $\boldsymbol{\tau}\cdot\operatorname{grad}_{\mathcal{S}}\chi_{\partial\mathcal{P}} = \mathbf{P}_{\!n}[\boldsymbol{\tau}]\cdot\operatorname{grad}_{\mathcal{S}}\chi_{\partial\mathcal{P}}$ while combining (423) and (424) through (422) along with the divergence theorem and the surface divergence theorem for nonsmooth closed surfaces (74), we are led to

$$\begin{aligned}&\int\limits_{\mathcal{P}} \chi_{\mathcal{P}}(\operatorname{div}\boldsymbol{\xi} + \pi + \gamma)\,\mathrm{d}v + \int\limits_{\partial\mathcal{P}} \chi_{\mathcal{P}}(\xi_{\mathcal{S}} - \boldsymbol{\xi}\cdot\boldsymbol{n})\,\mathrm{d}a \\ &\quad + \int\limits_{\partial\mathcal{P}} \chi_{\partial\mathcal{P}}(\operatorname{div}_{\mathcal{S}}(\mathbf{P}_{\!n}[\boldsymbol{\tau}]) + \varpi + \zeta - \xi_{\mathcal{S}})\,\mathrm{d}a + \int\limits_{\partial^2\mathcal{P}} \chi_{\partial\mathcal{P}}(\tau_{\partial\mathcal{S}} - \{\!\!\{\boldsymbol{\tau}\cdot\boldsymbol{\nu}\}\!\!\})\,\mathrm{d}\sigma = 0.\end{aligned} \tag{425}$$

Then, by variational arguments, the microtractions read

$$\xi_{\mathcal{S}} = \boldsymbol{\xi}\cdot\boldsymbol{n}, \qquad \text{and} \qquad \tau_{\partial\mathcal{S}} = \{\!\!\{\boldsymbol{\tau}\cdot\boldsymbol{\nu}\}\!\!\}, \tag{426}$$

while the bulk and surface field equations are given by

$$\operatorname{div}\boldsymbol{\xi} + \pi + \gamma = 0, \qquad \text{and} \qquad \operatorname{div}_{\mathcal{S}}(\mathbf{P}_{\!n}[\boldsymbol{\tau}]) + \varpi + \zeta - \xi_{\mathcal{S}} = 0. \tag{427}$$

Note that the bulk microforce balance $(427)_1$ has the standard form proposed by Fried & Gurtin [29]. However, the surface microforce balance $(427)_2$ has a contribution from the bulk, namely $\xi_{\mathcal{S}}$. Additionally, the term $\operatorname{div}_{\mathcal{S}}(\mathbf{P}_{\!n}[\boldsymbol{\tau}])$ may be split as $\operatorname{div}_{\mathcal{S}}(\mathbf{P}_{\!n}[\boldsymbol{\tau}]) = \operatorname{div}_{\mathcal{S}}\boldsymbol{\tau} + 2K\boldsymbol{\tau}\cdot\boldsymbol{n}$. Then, the surface microforce balance $(427)_2$ may be written as

$$\operatorname{div}_{\mathcal{S}}\boldsymbol{\tau} + 2K\boldsymbol{\tau}\cdot\boldsymbol{n} + \varpi + \zeta - \xi_{\mathcal{S}} = 0, \tag{428}$$

for each smooth part of $\partial\mathcal{P}$.

3 Conserved Species

We now account for the case where the bulk and surface phase fields, $\varphi_{\mathcal{P}}$ and $\varphi_{\partial\mathcal{P}}$, represent the concentration of a conserved species. We therefore supplement the field equations (427) by two partwise species balances, that is, the bulk species balance

$$\overline{\int_{\mathcal{P}} \varphi_{\mathcal{P}}\,\mathrm{d}v}^{\,\bullet} = \int_{\mathcal{P}} s_{\mathcal{P}}\,\mathrm{d}v - \int_{\partial\mathcal{P}} \boldsymbol{\jmath}_{\mathcal{P}}\cdot\boldsymbol{n}\,\mathrm{d}a, \tag{429}$$

and the surface species balance

$$\overline{\int_{\partial\mathcal{P}} \varphi_{\partial\mathcal{P}}\,\mathrm{d}a}^{\,\bullet} = \int_{\partial\mathcal{P}} \beta\,\boldsymbol{\jmath}_{\mathcal{P}}\cdot\boldsymbol{n}\,\mathrm{d}a + \int_{\partial\mathcal{P}} s_{\partial\mathcal{P}}\,\mathrm{d}a - \int_{\partial^2\mathcal{P}} \{\!\{\boldsymbol{\jmath}_{\partial\mathcal{P}}\cdot\boldsymbol{\nu}\}\!\}\,\mathrm{d}\sigma. \tag{430}$$

The partiwise bulk and surface balance of species, respectively, given by expressions (429) and (430), are motivated by the fact that we assume that the total balance of species satisfies

$$\overline{\int_{\mathcal{P}} \beta\,\varphi_{\mathcal{P}}\,\mathrm{d}v}^{\,\bullet} + \overline{\int_{\partial\mathcal{P}} \varphi_{\partial\mathcal{P}}\,\mathrm{d}a}^{\,\bullet} = \int_{\mathcal{P}} \beta\,s_{\mathcal{P}}\,\mathrm{d}v + \int_{\partial\mathcal{P}} s_{\partial\mathcal{P}}\,\mathrm{d}a - \int_{\partial^2\mathcal{P}} \{\!\{\boldsymbol{\jmath}_{\partial\mathcal{P}}\cdot\boldsymbol{\nu}\}\!\}\,\mathrm{d}\sigma. \tag{431}$$

Using the divergence theorem and the surface divergence theorem for nonsmooth closed surfaces (74) in expressions (429) and (430), respectively, followed by localization, we are led to

$$\dot{\varphi}_{\mathcal{P}} = s_{\mathcal{P}} - \mathrm{div}\,\boldsymbol{\jmath}_{\mathcal{P}}, \tag{432}$$

and

$$\dot{\varphi}_{\partial\mathcal{P}} = \beta\,\boldsymbol{\jmath}_{\mathcal{P}}\cdot\boldsymbol{n} + s_{\partial\mathcal{P}} - \mathrm{div}_{\mathcal{S}}(\mathbf{P}_{\!n}[\boldsymbol{\jmath}_{\partial\mathcal{P}}]). \tag{433}$$

Note that the bulk species balance has a standard form. However, the surface species balance has a contribution from the bulk, namely, $\beta\,\boldsymbol{\jmath}_{\mathcal{P}}\cdot\boldsymbol{n}$. Additionally, the term $\mathrm{div}_{\mathcal{S}}(\mathbf{P}_{\!n}[\boldsymbol{\jmath}_{\partial\mathcal{P}}])$ may be split as $\mathrm{div}_{\mathcal{S}}(\mathbf{P}_{\!n}[\boldsymbol{\jmath}_{\partial\mathcal{P}}]) = \mathrm{div}_{\mathcal{S}}\boldsymbol{\jmath}_{\partial\mathcal{P}} + 2K\boldsymbol{\jmath}_{\partial\mathcal{P}}\cdot\boldsymbol{n}$. Then, the surface species balance (433) may be written as

$$\dot{\varphi}_{\partial\mathcal{P}} = \beta\,\boldsymbol{\jmath}_{\mathcal{P}}\cdot\boldsymbol{n} + s_{\partial\mathcal{P}} - \mathrm{div}_{\mathcal{S}}\boldsymbol{\jmath}_{\partial\mathcal{P}} - 2K\boldsymbol{\jmath}_{\partial\mathcal{P}}\cdot\boldsymbol{n}, \tag{434}$$

for each smooth part of $\partial\mathcal{P}$.

4 Free-Energy Imbalance

First, note that the actual power is given by

$$\mathcal{W}_{\text{ext}}(\mathcal{P}, \partial\mathcal{P}) := \mathcal{V}_{\text{ext}}(\mathcal{P}, \partial\mathcal{P}; \dot{\varphi}_{\mathcal{P}}, \dot{\varphi}_{\partial\mathcal{P}}). \tag{435}$$

In the free-energy imbalance, together with the external power expenditure, we account for the rate at which energy is transferred to $\mathcal{P}$ and $\partial\mathcal{P}$ due to species transport. Thus, the free-energy imbalance reads

$$\begin{aligned}\overline{\int_{\mathcal{P}} \psi_{\mathcal{P}}\, \mathrm{d}v}^{\boldsymbol{\cdot}} + \overline{\int_{\partial\mathcal{P}} \psi_{\partial\mathcal{P}}\, \mathrm{d}a}^{\boldsymbol{\cdot}} \le \mathcal{W}_{\text{ext}}(\mathcal{P}, \partial\mathcal{P}) \\ + \int_{\mathcal{P}} \mu_{\mathcal{P}} s_{\mathcal{P}}\, \mathrm{d}v - \int_{\partial\mathcal{P}} \mu_{\mathcal{P}} \boldsymbol{\jmath}_{\mathcal{P}} \cdot \boldsymbol{n}\, \mathrm{d}a \\ + \int_{\partial\mathcal{P}} \beta\mu_{\partial\mathcal{P}} \boldsymbol{\jmath}_{\mathcal{P}} \cdot \boldsymbol{n}\, \mathrm{d}a + \int_{\partial\mathcal{P}} \mu_{\partial\mathcal{P}} s_{\partial\mathcal{P}}\, \mathrm{d}a - \int_{\partial^2\mathcal{P}} \{\!\{ \mu_{\partial\mathcal{P}} \boldsymbol{\jmath}_{\partial\mathcal{P}} \cdot \boldsymbol{\nu} \}\!\}\, \mathrm{d}\sigma.\end{aligned} \tag{436}$$

Noting that $\mathbf{P}_n[\boldsymbol{\jmath}_{\partial\mathcal{P}}] \cdot \operatorname{grad}_{\mathcal{S}} \mu_{\partial\mathcal{P}} = \boldsymbol{\jmath}_{\partial\mathcal{P}} \cdot \operatorname{grad}_{\mathcal{S}} \mu_{\partial\mathcal{P}}$ and uncoupling $\partial\mathcal{P}$ from $\mathcal{P}$, for the sake of simplicity, we have that

$$\dot{\psi}_{\mathcal{P}} + (\pi - \mu_{\mathcal{P}})\dot{\varphi}_{\mathcal{P}} - \boldsymbol{\xi} \cdot \operatorname{grad} \dot{\varphi}_{\mathcal{P}} + \boldsymbol{\jmath}_{\mathcal{P}} \cdot \operatorname{grad} \mu_{\mathcal{P}} \le 0, \tag{437}$$

and

$$\dot{\psi}_{\partial\mathcal{P}} + (\varpi - \mu_{\partial\mathcal{P}})\dot{\varphi}_{\partial\mathcal{P}} - \boldsymbol{\tau} \cdot \operatorname{grad}_{\mathcal{S}} \dot{\varphi}_{\partial\mathcal{P}} + \boldsymbol{\jmath}_{\partial\mathcal{P}} \cdot \operatorname{grad}_{\mathcal{S}} \mu_{\partial\mathcal{P}} \le 0. \tag{438}$$

Additionally, assuming that the bulk and surface free-energy densities $\psi_{\mathcal{P}}$ and $\psi_{\partial\mathcal{P}}$ are, respectively, given by constitutive response functions that are independent of $\mu_{\mathcal{P}}$, $\mu_{\partial\mathcal{P}}$, $\operatorname{grad} \mu_{\mathcal{P}}$, and $\operatorname{grad}_{\mathcal{S}} \mu_{\partial\mathcal{P}}$

$$\psi_{\mathcal{P}} := \psi_{\mathcal{P}}(\varphi_{\mathcal{P}}, \operatorname{grad} \varphi_{\mathcal{P}}), \qquad \text{and} \qquad \psi_{\partial\mathcal{P}} := \psi_{\partial\mathcal{P}}(\varphi_{\partial\mathcal{P}}, \operatorname{grad}_{\mathcal{S}} \varphi_{\partial\mathcal{P}}), \tag{439}$$

we have that

$$\dot{\psi}_{\mathcal{P}} = \partial_{\varphi_{\mathcal{P}}} \psi_{\mathcal{P}} \dot{\varphi}_{\mathcal{P}} + \partial_{\operatorname{grad} \varphi_{\mathcal{P}}} \psi_{\mathcal{P}} \overline{(\operatorname{grad} \varphi_{\mathcal{P}})}^{\boldsymbol{\cdot}}, \tag{440}$$

and

$$\dot{\psi}_{\partial\mathcal{P}} = \partial_{\varphi_{\partial\mathcal{P}}} \psi_{\partial\mathcal{P}} \dot{\varphi}_{\partial\mathcal{P}} + \partial_{\operatorname{grad}_{\mathcal{S}} \varphi_{\partial\mathcal{P}}} \psi_{\partial\mathcal{P}} \overline{(\operatorname{grad}_{\mathcal{S}} \varphi_{\partial\mathcal{P}})}^{\boldsymbol{\cdot}}. \tag{441}$$

Then, combining (437), (438), (440), and (441), we are led to two pointwise free-energy imbalances

$$(\mu_{\mathcal{P}} - \pi - \partial_{\varphi_{\mathcal{P}}} \psi_{\mathcal{P}})\dot{\varphi}_{\mathcal{P}} + (\boldsymbol{\xi} - \partial_{\operatorname{grad} \varphi_{\mathcal{P}}} \psi_{\mathcal{P}}) \cdot \operatorname{grad} \dot{\varphi}_{\mathcal{P}} - \boldsymbol{\jmath}_{\mathcal{P}} \cdot \operatorname{grad} \mu_{\mathcal{P}} \ge 0, \tag{442}$$

and

$$(\mu_{\partial\mathcal{P}} - \varpi - \partial_{\varphi_{\partial\mathcal{P}}}\psi_{\partial\mathcal{P}})\dot{\varphi}_{\partial\mathcal{P}} + (\boldsymbol{\tau} - \partial_{\mathrm{grad}_{\mathcal{S}}\varphi_{\partial\mathcal{P}}}\psi_{\partial\mathcal{P}}) \cdot \mathrm{grad}_{\mathcal{S}}\dot{\varphi}_{\partial\mathcal{P}} - \boldsymbol{\jmath}_{\partial\mathcal{P}} \cdot \mathrm{grad}_{\mathcal{S}}\mu_{\partial\mathcal{P}} \geq 0. \tag{443}$$

The bulk and surface free-energy imbalance (442) and (443) serve to devise additional constitutive response functions in what follows.

5 Additional Constitutive Response Functions

We now assume that the set of independent variables is given by

$$\{\varphi_{\mathcal{P}}, \varphi_{\partial\mathcal{P}}, \mathrm{grad}\,\varphi_{\mathcal{P}}, \mathrm{grad}_{\mathcal{S}}\varphi_{\partial\mathcal{P}}, \mu_{\mathcal{P}}, \mu_{\partial\mathcal{P}}\}, \tag{444}$$

while the set of dependent variables is

$$\{\pi, \varpi, \boldsymbol{\xi}, \boldsymbol{\tau}, \boldsymbol{\jmath}_{\mathcal{P}}, \boldsymbol{\jmath}_{\partial\mathcal{P}}\}. \tag{445}$$

Thus, we find that the local inequality (442) and (443) are satisfied in all processes if and only if:

- The bulk and surface microstress $\boldsymbol{\xi}$ and $\boldsymbol{\tau}$ are, respectively, given by

$$\boldsymbol{\xi} := \partial_{\mathrm{grad}\,\varphi_{\mathcal{P}}}\psi_{\mathcal{P}}, \qquad \text{and} \qquad \boldsymbol{\tau} := \partial_{\mathrm{grad}_{\mathcal{S}}\varphi_{\partial\mathcal{P}}}\psi_{\partial\mathcal{P}}. \tag{446}$$

- The internal bulk and surface microforces π and ϖ are, respectively, given by constitutive response functions that differ from the bulk and surface chemical potential by a contribution derived from the response functions $\psi_{\mathcal{P}}$ and $\psi_{\partial\mathcal{P}}$

$$\pi := \mu_{\mathcal{P}} - \partial_{\varphi_{\mathcal{P}}}\psi_{\mathcal{P}}, \qquad \text{and} \qquad \varpi := \mu_{\partial\mathcal{P}} - \partial_{\varphi_{\partial\mathcal{P}}}\psi_{\partial\mathcal{P}}. \tag{447}$$

- Granted that the bulk and surface species fluxes $\boldsymbol{\jmath}_{\mathcal{P}}$ and $\boldsymbol{\jmath}_{\partial\mathcal{P}}$ depend smoothly on the gradient of the bulk chemical potential, $\mathrm{grad}\,\mu_{\mathcal{P}}$, and the surface gradient of the surface chemical potentials, $\mathrm{grad}_{\mathcal{S}}\mu_{\partial\mathcal{P}}$, these fluxes are, respectively, given by a constitutive response function of the form

$$\boldsymbol{\jmath}_{\mathcal{P}} := -\boldsymbol{M}_{\mathcal{P}}[\mathrm{grad}\,\mu_{\mathcal{P}}], \qquad \text{and} \qquad \boldsymbol{\jmath}_{\partial\mathcal{P}} := -\boldsymbol{M}_{\partial\mathcal{P}}[\mathrm{grad}_{\mathcal{S}}\mu_{\partial\mathcal{P}}], \tag{448}$$

  where the mobility tensors $\boldsymbol{M}_{\mathcal{P}}$ and $\boldsymbol{M}_{\partial\mathcal{P}}$ must obey the residual dissipation inequalities

$$\mathrm{grad}\,\mu_{\mathcal{P}} \cdot \boldsymbol{M}_{\mathcal{P}}[\mathrm{grad}\,\mu_{\mathcal{P}}] \geq 0, \qquad \text{and} \qquad \mathrm{grad}_{\mathcal{S}}\mu_{\partial\mathcal{P}} \cdot \boldsymbol{M}_{\partial\mathcal{P}}[\mathrm{grad}_{\mathcal{S}}\mu_{\partial\mathcal{P}}] \geq 0. \tag{449}$$

For the sake of simplicity, we let $\boldsymbol{M}_{\mathcal{P}} := m_{\mathcal{P}}\mathbf{1}$ and $\boldsymbol{M}_{\partial\mathcal{P}} := m_{\partial\mathcal{P}}\mathbf{1}$. With this choice for $\boldsymbol{M}_{\partial\mathcal{P}}$, the surface flux $\boldsymbol{\jmath}_{\partial\mathcal{P}}$ remains proportional to $\mathrm{grad}_{\mathcal{S}}\mu_{\partial\mathcal{P}}$ and therefore tangential to $\partial\mathcal{P}$. Thus, the normal component of $\boldsymbol{\jmath}_{\partial\mathcal{P}}$ appearing in expression (434) vanishes. Also, note that nonlinear constitutive response functions for $\boldsymbol{\jmath}_{\mathcal{P}}$ and $\boldsymbol{\jmath}_{\partial\mathcal{P}}$ could be admissible as well, for instance, if $h\colon \mathbb{R}^3 \mapsto \mathbb{R}$ is a convex, differentiable function, one could assume $\boldsymbol{\jmath}_{\mathcal{P}} := \mathrm{grad}\, h(\mathrm{grad}\,\mu_{\mathcal{P}})$.

Important to what follows is the explicit form of the bulk and surface chemical potentials when using $(427)_1$ and (428) in (447). That is,

$$\mu_{\mathcal{P}} = -\mathrm{div}\,\boldsymbol{\xi} - \gamma + \partial_{\varphi_{\mathcal{P}}}\psi_{\mathcal{P}}, \tag{450}$$

and

$$\mu_{\partial\mathcal{P}} = -\mathrm{div}_{\mathcal{S}}\boldsymbol{\tau} - 2K\boldsymbol{\tau}\cdot\boldsymbol{n} - \zeta + \boldsymbol{\xi}\cdot\boldsymbol{n} + \partial_{\varphi_{\partial\mathcal{P}}}\psi_{\partial\mathcal{P}}, \tag{451}$$

which take the following form when considering (446),

$$\mu_{\mathcal{P}} = \partial_{\varphi_{\mathcal{P}}}\psi_{\mathcal{P}} - \mathrm{div}\left(\partial_{\mathrm{grad}\,\varphi_{\mathcal{P}}}\psi_{\mathcal{P}}\right) - \gamma, \tag{452}$$

and

$$\mu_{\partial\mathcal{P}} = \partial_{\varphi_{\partial\mathcal{P}}}\psi_{\partial\mathcal{P}} - \mathrm{div}_{\mathcal{S}}\left(\partial_{\mathrm{grad}_{\mathcal{S}}\varphi_{\partial\mathcal{P}}}\psi_{\partial\mathcal{P}}\right) - \zeta + \partial_{\mathrm{grad}\,\varphi_{\mathcal{P}}}\psi_{\mathcal{P}}\cdot\boldsymbol{n}, \tag{453}$$

for each smooth part of $\partial\mathcal{P}$. Note that the normal component of $\boldsymbol{\tau}$ vanishes for the free-energy function (414) with $(446)_2$.

In what follows, consider that $\mathcal{P} := \mathcal{B}$.

5.1 Further Connections: Boundary Conditions

To be slightly more general, let us define different parts of the boundary $\partial\mathcal{P}$. Let $\partial\mathcal{P}^{\mathrm{dyn}}$ be the boundary with the dynamic bulk-surface interplay and $\partial\mathcal{P}^{\mathrm{sta}}$ the static boundary such that $\partial\mathcal{P} := \partial\mathcal{P}^{\mathrm{dyn}} \cup \partial\mathcal{P}^{\mathrm{sta}}$ and $\partial\mathcal{P}^{\mathrm{dyn}} \cap \partial\mathcal{P}^{\mathrm{sta}} = \varnothing$. Also, let ${}_{\circ}\partial^2\mathcal{P}$ denote the boundary of the dynamic boundary $\partial\mathcal{P}^{\mathrm{dyn}}$ while $\partial^2\mathcal{P}$ still denotes the edge along which the normal field is discontinuous. Thus, given the boundary conditions derived based upon thermodynamical principles in [32, 33, 61], we stipulate that the boundary conditions may be prescribed as follows.

The essential boundary conditions, that is, the assignment of microstructure, read

$$\varphi_{\mathcal{P}}(\boldsymbol{x},t) = \varphi_{\partial\mathcal{P}}(\boldsymbol{x},t), \qquad \forall \boldsymbol{x} \in \partial\mathcal{P}^{\mathrm{dyn}}_{\mathrm{ess}}, \tag{454}$$

and

$$\varphi_{\partial\mathcal{P}}(\boldsymbol{x},t) = \varphi^{\mathrm{env}}_{\partial^2\mathcal{P}}(\boldsymbol{x}), \qquad \forall \boldsymbol{x} \in \partial^2\mathcal{P}^{\mathrm{sta}}_{\mathrm{ess}} \cup {}_{\circ}\partial^2\mathcal{P}^{\mathrm{sta}}_{\mathrm{ess}}, \tag{455}$$

where the surface phase field $\varphi_{\partial\mathcal{P}}$ is the action of the dynamic environment on $\partial\mathcal{P}^{\mathrm{dyn}}_{\mathrm{ess}}$ and $\varphi^{\mathrm{env}}_{\partial^2\mathcal{P}}$ is the action of the static environment on $\partial^2\mathcal{P}^{\mathrm{sta}}_{\mathrm{ess}}$. On a static environment, expression (454) takes the form

$$\varphi_{\mathcal{P}}(\boldsymbol{x},t) = \varphi^{\mathrm{env}}_{\partial\mathcal{P}}(\boldsymbol{x}), \qquad \forall \boldsymbol{x} \in \partial\mathcal{P}^{\mathrm{sta}}_{\mathrm{ess}}, \tag{456}$$

where $\varphi^{\mathrm{env}}_{\partial\mathcal{P}}$ is the action of the static environment on $\partial\mathcal{P}^{\mathrm{sta}}_{\mathrm{ess}}$.

Instead, we may opt for the natural boundary conditions, that is, the assignment of microtractions. Then, we have

$$\xi_{\mathcal{S}}(\boldsymbol{x},t) = \xi^{\mathrm{env}}_{\mathcal{S}}(\boldsymbol{x}), \qquad \forall \boldsymbol{x} \in \partial\mathcal{P}^{\mathrm{sta}}_{\mathrm{nat}}, \tag{457}$$

$$\tau_{\partial\mathcal{S}}(\boldsymbol{x},t) = \tau^{\mathrm{env}}_{\partial\mathcal{S}}(\boldsymbol{x}), \qquad \forall \boldsymbol{x} \in \partial^2\mathcal{P}^{\mathrm{sta}}_{\mathrm{nat}}, \tag{458}$$

and

$$_{\circ}\tau_{\partial\mathcal{S}}(\boldsymbol{x},t) = \tau^{\mathrm{env}}_{\partial\mathcal{S}}(\boldsymbol{x}), \qquad \forall \boldsymbol{x} \in {}_{\circ}\partial^2\mathcal{P}^{\mathrm{sta}}_{\mathrm{ess}}, \tag{459}$$

where $\xi_{\mathcal{S}} = \boldsymbol{\xi}\cdot\boldsymbol{n}$, $\tau_{\partial\mathcal{S}} = \{\!\{\boldsymbol{\tau}\cdot\boldsymbol{\nu}\}\!\}$ on $\partial^2\mathcal{P}^{\mathrm{sta}}_{\mathrm{nat}}$, ${}_{\circ}\tau_{\partial\mathcal{S}} = \boldsymbol{\tau}\cdot\boldsymbol{\nu}$ on ${}_{\circ}\partial^2\mathcal{P}^{\mathrm{sta}}_{\mathrm{nat}}$ due to (77), and $\xi^{\mathrm{env}}_{\mathcal{S}}$ and $\tau^{\mathrm{env}}_{\partial\mathcal{S}}$ are the actions of the static environment, respectively, on $\partial\mathcal{P}^{\mathrm{sta}}_{\mathrm{nat}}$ and $\partial^2\mathcal{P}^{\mathrm{sta}}_{\mathrm{nat}} \cup {}_{\circ}\partial^2\mathcal{P}^{\mathrm{sta}}_{\mathrm{nat}}$.

As for the bulk species balance, as an essential boundary condition, we may prescribe

$$\mu_{\mathcal{P}}(\boldsymbol{x},t) = \beta\mu_{\partial\mathcal{P}}(\boldsymbol{x},t), \qquad \forall \boldsymbol{x} \in \partial\mathcal{P}^{\mathrm{dyn}}_{\mathrm{ess}}, \tag{460}$$

where $\beta\mu_{\partial\mathcal{P}}$ is the action of the dynamic environment on $\partial\mathcal{P}^{\mathrm{sta}}_{\mathrm{nat}}$. Expression (460), on a static environment, takes the form

$$\mu_{\mathcal{P}}(\boldsymbol{x},t) = \mu^{\mathrm{env}}_{\partial\mathcal{P}}(\boldsymbol{x}), \qquad \forall \boldsymbol{x} \in \partial\mathcal{P}^{\mathrm{sta}}_{\mathrm{ess}}, \tag{461}$$

where $\mu^{\mathrm{env}}_{\partial\mathcal{P}}$ is the action of the static environment on $\partial\mathcal{P}^{\mathrm{sta}}_{\mathrm{ess}}$.

Instead, as a natural boundary condition, we may opt for

$$\boldsymbol{\jmath}_{\mathcal{P}}(\boldsymbol{x},t)\cdot\boldsymbol{n} = -\jmath^{\mathrm{env}}_{\partial\mathcal{P}}(\boldsymbol{x}), \qquad \forall \boldsymbol{x} \in \partial\mathcal{P}^{\mathrm{sta}}_{\mathrm{nat}}, \tag{462}$$

where $\jmath^{\mathrm{env}}_{\partial\mathcal{P}}$ is the action of the static environment on $\partial\mathcal{P}^{\mathrm{sta}}_{\mathrm{nat}}$ through the normal component of the flux $\boldsymbol{\jmath}_{\mathcal{P}}$. Conversely, for the surface species balance, as an essential boundary condition, we may prescribe

$$\mu_{\partial\mathcal{P}}(\boldsymbol{x},t) = \mu^{\mathrm{env}}_{\partial^2\mathcal{P}}(\boldsymbol{x}), \qquad \forall \boldsymbol{x} \in \partial^2\mathcal{P}^{\mathrm{sta}}_{\mathrm{ess}} \cup {}_{\circ}\partial^2\mathcal{P}^{\mathrm{sta}}_{\mathrm{ess}}, \tag{463}$$

where $\mu^{\mathrm{env}}_{\partial^2\mathcal{P}}$ is the action of the static environment on $\partial^2\mathcal{P}^{\mathrm{sta}}_{\mathrm{ess}} \cup {}_{\circ}\partial^2\mathcal{P}^{\mathrm{sta}}_{\mathrm{ess}}$, whereas, as a natural boundary condition, we may opt for

$$\{\!\{\boldsymbol{\jmath}_{\partial\mathcal{P}}(\boldsymbol{x},t)\cdot\boldsymbol{\nu}\}\!\} = -\jmath^{\mathrm{env}}_{\partial^2\mathcal{P}}(\boldsymbol{x}), \qquad \forall \boldsymbol{x} \in \partial^2\mathcal{P}^{\mathrm{sta}}_{\mathrm{nat}}, \tag{464}$$

and, due to (77),

$$\jmath_{\partial\mathcal{P}}(\boldsymbol{x},t)\cdot\boldsymbol{\nu} = -\jmath^{\text{env}}_{\partial^2\mathcal{P}}(\boldsymbol{x}), \qquad \forall \boldsymbol{x}\in {}_{\circ}\partial^2\mathcal{P}^{\text{sta}}_{\text{nat}}, \tag{465}$$

where $\jmath^{\text{env}}_{\partial^2\mathcal{P}}$ is the action of the static environment on $\partial^2\mathcal{P}^{\text{sta}}_{\text{nat}}$ through the tangent-normal component of the flux $\jmath_{\partial\mathcal{P}}$.

Now, we restrict attention to the mixed boundary conditions on $\partial\mathcal{P}^{\text{dyn}}$. Specifically, we invoke relation proposed by Fried & Gurtin [20, surface free-energy imbalance (92)] and stipulate that

$$\mathcal{T}_{\text{surf}}(-\partial\mathcal{P}) + \mathcal{T}_{\text{env}}(\partial\mathcal{P}) \geq 0, \tag{466}$$

where $\mathcal{T}_{\text{surf}}(-\partial\mathcal{P})$ combines the power expended on $\partial\mathcal{P}$ by the material inside $\mathcal{P}$ and the rate at which energy is transferred to $\mathcal{P}$ and $\partial\mathcal{P}$, whereas $\mathcal{T}_{\text{env}}(\partial\mathcal{P})$ combines the power expended by the environment on $\partial\mathcal{P}$ and the rate at which energy is transferred from the environment to $\partial\mathcal{P}$. For $\partial\mathcal{P}^{\text{dyn}}$, we here define

$$\begin{aligned}\mathcal{T}_{\text{surf}}(-\partial\mathcal{P}^{\text{dyn}}) := &-\int_{\partial\mathcal{P}} (\dot{\varphi}_{\mathcal{P}} - \dot{\varphi}_{\partial\mathcal{P}})\xi_{\mathcal{S}}\,\mathrm{d}a - \int_{\partial^2\mathcal{P}} \tau_{\partial\mathcal{S}}\dot{\varphi}_{\partial\mathcal{P}}\,\mathrm{d}\sigma + \int_{{}_{\circ}\partial^2\mathcal{P}} {}_{\circ}\tau_{\partial\mathcal{S}}\dot{\varphi}_{\partial\mathcal{P}}\,\mathrm{d}\sigma \\ &-\int_{\partial\mathcal{P}} (\beta\mu_{\partial\mathcal{P}} - \mu_{\mathcal{P}})\jmath_{\mathcal{P}}\cdot\boldsymbol{n}\,\mathrm{d}a + \int_{\partial^2\mathcal{P}} \{\!\{\, \mu_{\partial\mathcal{P}}\jmath_{\partial\mathcal{P}}\cdot\boldsymbol{\nu} \,\}\!\}\,\mathrm{d}\sigma \\ &+\int_{{}_{\circ}\partial^2\mathcal{P}} \mu_{\partial\mathcal{P}}\jmath_{\partial\mathcal{P}}\cdot\boldsymbol{\nu}\,\mathrm{d}\sigma,\end{aligned} \tag{467}$$

where, owing to (77), ${}_{\circ}\tau_{\partial\mathcal{S}} := \boldsymbol{\tau}\cdot\boldsymbol{\nu}$ is the analogous of $\tau_{\partial\mathcal{S}}$ but developed on ${}_{\circ}\partial^2\mathcal{P}$. We also define

$$\begin{aligned}\mathcal{T}_{\text{env}}(\partial\mathcal{P}^{\text{dyn}}) := &\int_{\partial^2\mathcal{P}} \tau^{\text{env}}_{\partial\mathcal{S}}\dot{\varphi}^{\text{env}}_{\partial^2\mathcal{P}}\,\mathrm{d}\sigma + \int_{{}_{\circ}\partial^2\mathcal{P}} {}_{\circ}\tau^{\text{env}}_{\partial\mathcal{S}}\dot{\varphi}^{\text{env}}_{\partial^2\mathcal{P}}\,\mathrm{d}\sigma \\ &-\int_{\partial^2\mathcal{P}} \mu^{\text{env}}_{\partial^2\mathcal{P}}\jmath^{\text{env}}_{\partial^2\mathcal{P}}\,\mathrm{d}\sigma - \int_{{}_{\circ}\partial^2\mathcal{P}} \mu^{\text{env}}_{\partial^2\mathcal{P}}\jmath^{\text{env}}_{\partial^2\mathcal{P}}\,\mathrm{d}\sigma.\end{aligned} \tag{468}$$

Now, we consider that on $\partial^2\mathcal{P}\cup{}_{\circ}\partial^2\mathcal{P}$, $\tau^{\text{env}}_{\partial\mathcal{S}} = \tau_{\partial\mathcal{S}}$, $\dot{\varphi}^{\text{env}}_{\partial^2\mathcal{P}} = \dot{\varphi}_{\partial\mathcal{P}}$, $\mu^{\text{env}}_{\partial^2\mathcal{P}} = \mu_{\partial\mathcal{P}}$, whereas on $\partial^2\mathcal{P}$, $\jmath^{\text{env}}_{\partial^2\mathcal{P}} = \{\!\{\, \jmath_{\partial\mathcal{P}}\cdot\boldsymbol{\nu} \,\}\!\}$ and on ${}_{\circ}\partial^2\mathcal{P}$, $\jmath^{\text{env}}_{\partial^2\mathcal{P}} = \jmath_{\partial\mathcal{P}}\cdot\boldsymbol{\nu}$. Thus, expression (466) reads

$$-\int_{\partial\mathcal{P}} \big((\dot{\varphi}_{\mathcal{P}} - \dot{\varphi}_{\partial\mathcal{P}})\,\xi_{\mathcal{S}} + (\beta\mu_{\partial\mathcal{P}} - \mu_{\mathcal{P}})\,\jmath_{\mathcal{P}}\cdot\boldsymbol{n}\big)\,\mathrm{d}a \geq 0. \tag{469}$$

Uncoupling this expression, we have that

$$\int_{\partial\mathcal{P}} (\dot{\varphi}_{\mathcal{P}} - \dot{\varphi}_{\partial\mathcal{P}})\,\xi_{\mathcal{S}}\,\mathrm{d}a \leq 0, \qquad \text{and} \qquad \int_{\partial\mathcal{P}} (\beta\mu_{\partial\mathcal{P}} - \mu_{\mathcal{P}})\,\jmath_{\mathcal{P}} \cdot \boldsymbol{n}\,\mathrm{d}a \leq 0. \tag{470}$$

Note that, the terms in (470) are dissipative. That is, as a mixed boundary condition, expressions (470) read

$$\boldsymbol{\xi}(\boldsymbol{x}, t) \cdot \boldsymbol{n} = \tfrac{1}{L_\varphi}(\dot{\varphi}_{\partial\mathcal{P}} - \dot{\varphi}_{\mathcal{P}}), \qquad \forall \boldsymbol{x} \in \partial\mathcal{P}^{\mathrm{dyn}}_{\mathrm{mix}}, \tag{471}$$

and

$$\jmath_{\mathcal{P}}(\boldsymbol{x}, t) \cdot \boldsymbol{n} = -\tfrac{1}{L_\mu}(\beta\mu_{\partial\mathcal{P}} - \mu_{\mathcal{P}}), \qquad \forall \boldsymbol{x} \in \partial\mathcal{P}^{\mathrm{dyn}}_{\mathrm{mix}}, \tag{472}$$

where $L_\varphi, L_\mu > 0$.

5.2 Specialized Equations

In view of the free-energy functional (414), and expressions (432), (434), (452), and (453), our theory renders the following set of equations

$$\begin{cases} \dot{\varphi}_{\mathcal{P}} = s_{\mathcal{P}} + m_{\mathcal{P}}\Delta\mu_{\mathcal{P}}, & \text{in } \mathcal{P} \\ \mu_{\mathcal{P}} = -\epsilon\Delta\varphi_{\mathcal{P}} + \tfrac{1}{\epsilon}f'(\varphi_{\mathcal{P}}) - \gamma, & \text{in } \mathcal{P}, \\ \dot{\varphi}_{\partial\mathcal{P}} = s_{\partial\mathcal{P}} + m_{\partial\mathcal{P}}\Delta_{\mathcal{S}}\mu_{\partial\mathcal{P}} - \beta m_{\mathcal{P}}\,\partial_n\mu_{\mathcal{P}}, & \text{on } \partial\mathcal{P}, \\ \mu_{\partial\mathcal{P}} = -\iota\delta\Delta_{\mathcal{S}}\varphi_{\partial\mathcal{P}} + \tfrac{1}{\delta}g'(\varphi_{\partial\mathcal{P}}) + \epsilon\,\partial_n\varphi_{\mathcal{P}} - \zeta & \text{on } \partial\mathcal{P}. \end{cases} \tag{473}$$

The boundary conditions may be summarized as follows:

$$\forall \boldsymbol{x} \in \partial\mathcal{P}^{\mathrm{dyn}} \begin{cases} \varphi_{\mathcal{P}} = \varphi_{\partial\mathcal{P}}, & \text{or} \quad \epsilon\,\partial_n\varphi_{\mathcal{P}} = \tfrac{1}{L_\varphi}(\dot{\varphi}_{\partial\mathcal{P}} - \dot{\varphi}_{\mathcal{P}}), \\ \mu_{\mathcal{P}} = \beta\mu_{\partial\mathcal{P}}, & \text{or} \quad -m_{\mathcal{P}}\,\partial_n\mu_{\mathcal{P}} = -\tfrac{1}{L_\mu}(\beta\mu_{\partial\mathcal{P}} - \mu_{\mathcal{P}}), \end{cases} \tag{474}$$

and

$$\forall \boldsymbol{x} \in \partial\mathcal{P}^{\mathrm{sta}} \begin{cases} \varphi_{\mathcal{P}} = \varphi^{\mathrm{env}}_{\partial\mathcal{P}}, & \text{or} \quad \epsilon\,\partial_n\varphi_{\mathcal{P}} = \xi^{\mathrm{env}}_{\mathcal{S}}, \\ \mu_{\mathcal{P}} = \mu^{\mathrm{env}}_{\partial\mathcal{P}}, & \text{or} \quad -m_{\mathcal{P}}\,\partial_n\mu_{\mathcal{P}} = -\jmath^{\mathrm{env}}_{\partial\mathcal{P}}, \end{cases} \tag{475}$$

and

$$\forall \boldsymbol{x} \in \partial^2\mathcal{P}^{\mathrm{sta}} \begin{cases} \varphi_{\partial\mathcal{P}} = \varphi^{\mathrm{env}}_{\partial^2\mathcal{P}}, & \text{or} \quad \iota\delta\,\{\!\{\,\partial_\nu\varphi_{\partial\mathcal{P}}\,\}\!\} = \tau^{\mathrm{env}}_{\partial\mathcal{S}}, \\ \mu_{\partial\mathcal{P}} = \mu^{\mathrm{env}}_{\partial^2\mathcal{P}}, & \text{or} \quad -m_{\partial\mathcal{P}}\{\!\{\,\partial_\nu\mu_{\partial\mathcal{P}}\,\}\!\} = -\jmath^{\mathrm{env}}_{\partial^2\mathcal{P}}, \end{cases} \tag{476}$$

and

$$\forall \boldsymbol{x} \in {}_{\circ}\partial^2\mathcal{P}^{\mathrm{sta}} \begin{cases} \varphi_{\partial\mathcal{P}} = \varphi^{\mathrm{env}}_{\partial^2\mathcal{P}}, & \text{or} \quad \iota\delta\,\partial_\nu\varphi_{\partial\mathcal{P}} = \tau^{\mathrm{env}}_{\partial\mathcal{S}}, \\ \mu_{\partial\mathcal{P}} = \mu^{\mathrm{env}}_{\partial^2\mathcal{P}}, & \text{or} \quad -m_{\partial\mathcal{P}}\,\partial_\nu\mu_{\partial\mathcal{P}} = -\jmath^{\mathrm{env}}_{\partial^2\mathcal{P}}. \end{cases} \tag{477}$$

6 Decay Relations

We now aim to establish Lyapunov decay relations for the case where all the boundary $\partial\mathcal{P}$ is dynamic and of the mixed type for both the microstructure and chemical potential. Thus, in view of (440) and (441) combined with the constitutive relations for the bulk and surface microstresses (446), we have that

$$\begin{aligned}\int_{\mathcal{P}} \dot{\psi}_{\mathcal{P}}\,\mathrm{d}v + \int_{\partial\mathcal{P}} \dot{\psi}_{\partial\mathcal{P}}\,\mathrm{d}a &= \int_{\mathcal{P}} \big(\partial_{\varphi_{\mathcal{P}}}\psi_{\mathcal{P}}\dot{\varphi}_{\mathcal{P}} + \partial_{\mathrm{grad}\,\varphi_{\mathcal{P}}}\psi_{\mathcal{P}}(\mathrm{grad}\,\varphi_{\mathcal{P}})\dot{}\,\big)\,\mathrm{d}v \\ &\quad + \int_{\partial\mathcal{P}} \big(\partial_{\varphi_{\partial\mathcal{P}}}\psi_{\partial\mathcal{P}}\dot{\varphi}_{\partial\mathcal{P}} + \partial_{\mathrm{grad}_{\mathcal{S}}\varphi_{\partial\mathcal{P}}}\psi_{\partial\mathcal{P}}(\mathrm{grad}_{\mathcal{S}}\varphi_{\partial\mathcal{P}})\dot{}\,\big)\,\mathrm{d}a, \\ &= \int_{\mathcal{P}} \big(\partial_{\varphi_{\mathcal{P}}}\psi_{\mathcal{P}}\dot{\varphi}_{\mathcal{P}} + \boldsymbol{\xi}\cdot\mathrm{grad}\,\dot{\varphi}_{\mathcal{P}}\big)\,\mathrm{d}v \\ &\quad + \int_{\partial\mathcal{P}} \big(\partial_{\varphi_{\partial\mathcal{P}}}\psi_{\partial\mathcal{P}}\dot{\varphi}_{\partial\mathcal{P}} + \mathbf{P}_{\!n}[\boldsymbol{\tau}]\cdot\mathrm{grad}_{\mathcal{S}}\dot{\varphi}_{\partial\mathcal{P}}\big)\,\mathrm{d}a, \\ &= \int_{\mathcal{P}} \big((\partial_{\varphi_{\mathcal{P}}}\psi_{\mathcal{P}} - \mathrm{div}\,\boldsymbol{\xi})\dot{\varphi}_{\mathcal{P}} + \mathrm{div}\,(\dot{\varphi}_{\mathcal{P}}\,\boldsymbol{\xi})\big)\,\mathrm{d}v \\ &\quad + \int_{\partial\mathcal{P}} \big((\partial_{\varphi_{\partial\mathcal{P}}}\psi_{\partial\mathcal{P}} - \mathrm{div}_{\mathcal{S}}(\mathbf{P}_{\!n}[\boldsymbol{\tau}]))\dot{\varphi}_{\partial\mathcal{P}} + \mathrm{div}_{\mathcal{S}}(\dot{\varphi}_{\partial\mathcal{P}}\,\mathbf{P}_{\!n}[\boldsymbol{\tau}])\big)\,\mathrm{d}a.\end{aligned} \tag{478}$$

Next, using the pointwise balances of microforces (427), we arrive at

$$\begin{aligned}\int_{\mathcal{P}} \dot{\psi}_{\mathcal{P}}\,\mathrm{d}v + \int_{\partial\mathcal{P}} \dot{\psi}_{\partial\mathcal{P}}\,\mathrm{d}a &= \int_{\mathcal{P}} \big((\partial_{\varphi_{\mathcal{P}}}\psi_{\mathcal{P}} + \pi + \gamma)\dot{\varphi}_{\mathcal{P}} + \mathrm{div}\,(\dot{\varphi}_{\mathcal{P}}\,\boldsymbol{\xi})\big)\,\mathrm{d}v \\ &\quad + \int_{\partial\mathcal{P}} \big((\partial_{\varphi_{\partial\mathcal{P}}}\psi_{\partial\mathcal{P}} + \varpi + \zeta - \xi_{\mathcal{S}})\dot{\varphi}_{\partial\mathcal{P}} + \mathrm{div}_{\mathcal{S}}(\dot{\varphi}_{\partial\mathcal{P}}\,\mathbf{P}_{\!n}[\boldsymbol{\tau}])\big)\,\mathrm{d}a.\end{aligned} \tag{479}$$

Note that relation (479) holds whether or not the species transports in the bulk and on the surface are present or not. Conversely, in view of expression $(426)_1$, the surface microtraction, and accounting for the bulk and surface internal microforces, given by expressions (447), in (479), we are led to

$$\int_{\mathcal{P}} \dot{\psi}_{\mathcal{P}} \, \mathrm{d}v + \int_{\partial\mathcal{P}} \dot{\psi}_{\partial\mathcal{P}} \, \mathrm{d}a = \int_{\mathcal{P}} \big((\mu_{\mathcal{P}} + \gamma)\dot{\varphi}_{\mathcal{P}} + \operatorname{div}(\dot{\varphi}_{\mathcal{P}}\,\boldsymbol{\xi})\big)\, \mathrm{d}v + \int_{\partial\mathcal{P}} \big((\mu_{\partial\mathcal{P}} + \zeta - \boldsymbol{\xi}\cdot\boldsymbol{n})\dot{\varphi}_{\partial\mathcal{P}} + \operatorname{div}_{\mathcal{S}}(\dot{\varphi}_{\partial\mathcal{P}}\,\mathbf{P}_{\!n}[\boldsymbol{\tau}])\big)\, \mathrm{d}a. \tag{480}$$

We now use the bulk and surface species balances, respectively, given by (432) and (433) in expression (480), to arrive at

$$\begin{aligned}
\int_{\mathcal{P}} \dot{\psi}_{\mathcal{P}} \, \mathrm{d}v + \int_{\partial\mathcal{P}} \dot{\psi}_{\partial\mathcal{P}} \, \mathrm{d}a = {} & \\
& - \int_{\mathcal{P}} \operatorname{grad}\mu_{\mathcal{P}} \cdot \boldsymbol{M}_{\mathcal{P}}[\operatorname{grad}\mu_{\mathcal{P}}]\, \mathrm{d}v \\
& + \int_{\mathcal{P}} \big(\mu_{\mathcal{P}} s_{\mathcal{P}} + \gamma\dot{\varphi}_{\mathcal{P}} + \operatorname{div}(\dot{\varphi}_{\mathcal{P}}\,\boldsymbol{\xi} - \mu_{\mathcal{P}}\boldsymbol{\jmath}_{\mathcal{P}})\big)\, \mathrm{d}v \\
& - \int_{\partial\mathcal{P}} \operatorname{grad}_{\mathcal{S}}\mu_{\partial\mathcal{P}} \cdot \boldsymbol{M}_{\partial\mathcal{P}}[\operatorname{grad}_{\mathcal{S}}\mu_{\partial\mathcal{P}}]\, \mathrm{d}a \\
& + \int_{\partial\mathcal{P}} \big(\mu_{\partial\mathcal{P}}(\beta\,\boldsymbol{\jmath}_{\mathcal{P}}\cdot\boldsymbol{n} + s_{\partial\mathcal{P}})\big)\, \mathrm{d}a \\
& + \int_{\partial\mathcal{P}} \big((\zeta - \boldsymbol{\xi}\cdot\boldsymbol{n})\dot{\varphi}_{\partial\mathcal{P}} + \operatorname{div}_{\mathcal{S}}(\dot{\varphi}_{\partial\mathcal{P}}\,\mathbf{P}_{\!n}[\boldsymbol{\tau}] - \mu_{\partial\mathcal{P}}\,\mathbf{P}_{\!n}[\boldsymbol{\jmath}_{\partial\mathcal{P}}])\big)\, \mathrm{d}a.
\end{aligned} \tag{481}$$

Then, using the divergence theorem and the surface divergence theorem for closed nonsmooth surfaces (74), we are led to

$$\begin{aligned}
\int_{\mathcal{P}} \dot{\psi}_{\mathcal{P}} \, \mathrm{d}v + \int_{\partial\mathcal{P}} \dot{\psi}_{\partial\mathcal{P}} \, \mathrm{d}a = {} & - \int_{\mathcal{P}} \operatorname{grad}\mu_{\mathcal{P}} \cdot \boldsymbol{M}_{\mathcal{P}}[\operatorname{grad}\mu_{\mathcal{P}}]\, \mathrm{d}v \\
& + \int_{\mathcal{P}} \big(\mu_{\mathcal{P}} s_{\mathcal{P}} + \gamma\dot{\varphi}_{\mathcal{P}}\big)\, \mathrm{d}v \\
& - \int_{\partial\mathcal{P}} \operatorname{grad}_{\mathcal{S}}\mu_{\partial\mathcal{P}} \cdot \boldsymbol{M}_{\partial\mathcal{P}}[\operatorname{grad}_{\mathcal{S}}\mu_{\partial\mathcal{P}}]\, \mathrm{d}a \\
& + \int_{\partial\mathcal{P}} \big(\mu_{\partial\mathcal{P}} s_{\partial\mathcal{P}} + (\beta\mu_{\partial\mathcal{P}} - \mu_{\mathcal{P}})\,\boldsymbol{\jmath}_{\mathcal{P}}\cdot\boldsymbol{n}\big)\, \mathrm{d}a \\
& + \int_{\partial\mathcal{P}} \big(\zeta\dot{\varphi}_{\partial\mathcal{P}} + (\dot{\varphi}_{\mathcal{P}} - \dot{\varphi}_{\partial\mathcal{P}})\,\boldsymbol{\xi}\cdot\boldsymbol{n}\big)\, \mathrm{d}a
\end{aligned}$$

$$+ \int\limits_{\partial^2\mathcal{P}} \left(\{\!\{ \dot{\varphi}_{\partial\mathcal{P}} \, \boldsymbol{\tau} \cdot \boldsymbol{\nu} \}\!\} - \{\!\{ \mu_{\partial\mathcal{P}} \, \boldsymbol{\jmath}_{\partial\mathcal{P}} \cdot \boldsymbol{\nu} \}\!\} \right) \mathrm{d}\sigma. \tag{482}$$

6.1 Decay Relations for Mixed Boundary Conditions

For a mixed boundary condition on $\partial\mathcal{P} := \partial\mathcal{P}^{\mathrm{dyn}}_{\mathrm{mix}}$ given by expressions (471) and (472) along with the boundary-edge conditions (458) and (464), we obtained the final relation

$$\begin{aligned}
&\int\limits_{\mathcal{P}} \dot{\psi}_{\mathcal{P}} \, \mathrm{d}v + \int\limits_{\partial\mathcal{P}} \dot{\psi}_{\partial\mathcal{P}} \, \mathrm{d}a = \\
&\int\limits_{\mathcal{P}} \left(-\operatorname{grad}\mu_{\mathcal{P}} \cdot \boldsymbol{M}_{\mathcal{P}}[\operatorname{grad}\mu_{\mathcal{P}}] + \mu_{\mathcal{P}} s_{\mathcal{P}} + \gamma \dot{\varphi}_{\mathcal{P}} \right) \mathrm{d}v \\
&+ \int\limits_{\partial\mathcal{P}} \left(-\operatorname{grad}_{\mathcal{S}}\mu_{\partial\mathcal{P}} \cdot \boldsymbol{M}_{\partial\mathcal{P}}[\operatorname{grad}_{\mathcal{S}}\mu_{\partial\mathcal{P}}] + \mu_{\partial\mathcal{P}} s_{\partial\mathcal{P}} + \zeta \dot{\varphi}_{\partial\mathcal{P}} \right) \mathrm{d}a \\
&+ \int\limits_{\partial\mathcal{P}} \left(-\tfrac{1}{L_{\varphi}} (\dot{\varphi}_{\partial\mathcal{P}} - \dot{\varphi}_{\mathcal{P}})^2 - \tfrac{1}{L_{\mu}} (\beta\mu_{\partial\mathcal{P}} - \mu_{\mathcal{P}})^2 \right) \mathrm{d}a \\
&+ \int\limits_{\partial^2\mathcal{P}} \left(\dot{\varphi}_{\partial\mathcal{P}} \, \tau^{\mathrm{env}}_{\partial\mathcal{S}} + \mu_{\partial\mathcal{P}} \, \jmath^{\mathrm{env}}_{\partial^2\mathcal{P}} \right) \mathrm{d}\sigma.
\end{aligned} \tag{483}$$

Thus, the final Lyapunov decay relation is

$$\begin{aligned}
\int\limits_{\mathcal{P}} \dot{\psi}_{\mathcal{P}} \, \mathrm{d}v + \int\limits_{\partial\mathcal{P}} \dot{\psi}_{\partial\mathcal{P}} \, \mathrm{d}a \leq \int\limits_{\mathcal{P}} \left(\mu_{\mathcal{P}} s_{\mathcal{P}} + \gamma \dot{\varphi}_{\mathcal{P}} \right) \mathrm{d}v + \int\limits_{\partial\mathcal{P}} \left(\mu_{\partial\mathcal{P}} s_{\partial\mathcal{P}} + \zeta \dot{\varphi}_{\partial\mathcal{P}} \right) \mathrm{d}a \\
+ \int\limits_{\partial^2\mathcal{P}} \left(\dot{\varphi}_{\partial\mathcal{P}} \, \tau^{\mathrm{env}}_{\partial\mathcal{S}} + \mu_{\partial\mathcal{P}} \, \jmath^{\mathrm{env}}_{\partial^2\mathcal{P}} \right) \mathrm{d}\sigma.
\end{aligned} \tag{484}$$

Part VI
Conclusion

Chapter 10
Summary and Specializations

To compare and understand the different implications of considering different level of arbitrariness, we summarize our results and specialize them. First, In Fig. 1, we depict the body under consideration. In what follows, we take three different arbitrary parts to summarize results for first, second, and third gradient theories, respectively, when motion and a single phase field are considered. For the summary of the first and second gradient theories, we also referred the reader to the works by Gurtin et al. [34], and Fried & Gurtin [20]. For the first gradient theory, the arbitrary part is enclosed by a smooth surface, while for the second gradient theory, we allow the enclosing surface to lack smoothness at a junction-edge; however, the junction-edge is smooth. For the third gradient theory, we also allow the junction-edge to lack smoothness at a point.

L. Espath, *Mechanics and Geometry of Enriched Continua*,
https://doi.org/10.1007/978-3-031-28934-7_10

Fig. 1 Body $\mathcal{B}$

—Summary of a first gradient theory—

Incompressible fluid flow with one conserved phase field:

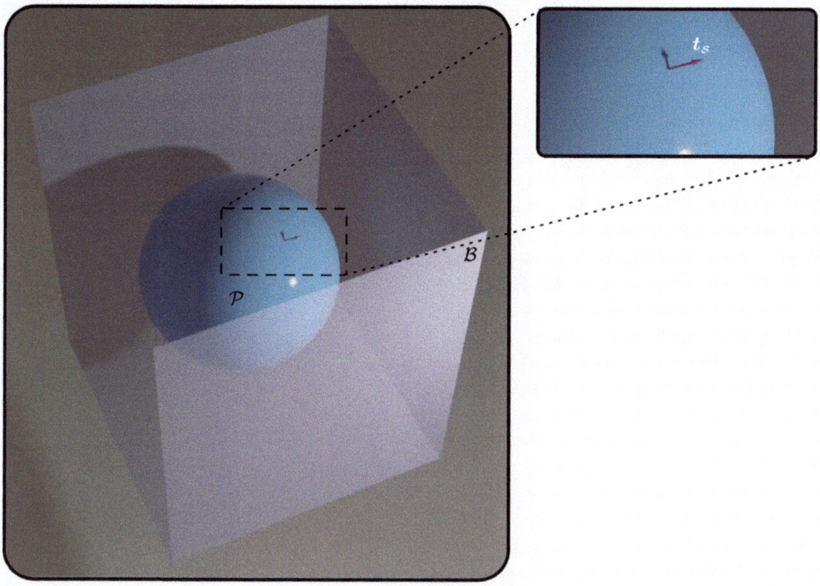

An arbitrary part of $\mathcal{B}$

- *Microtractions, tractions, couple-microtractions, and couple-tractions*

$$\xi_S = \boldsymbol{\xi} \cdot \boldsymbol{n}$$

$$\boldsymbol{t}_S = \mathbf{T}[\boldsymbol{n}]$$

- *Free-energy density*

$$\psi = f(\varphi) + \tfrac{1}{2}\|\boldsymbol{y}\|^2$$

f: potential function, $\boldsymbol{y} = \operatorname{grad}\varphi$, $\mathbf{B} := \operatorname{sym}(\operatorname{grad}\boldsymbol{v})$

- *Pointwise field equations*

$$\operatorname{div}(\mathbf{S} - \mathbf{1}g - \operatorname{grad}\varphi \otimes \boldsymbol{\xi}) + \boldsymbol{b}^{\mathrm{ni}} - \varrho\dot{\boldsymbol{v}} = \mathbf{0}$$

$$\operatorname{div}\partial_{\operatorname{grad}\varphi}\psi - \partial_{\varphi}\psi + \gamma/\varrho = -\mu$$

$$\operatorname{div}(\mathbf{M}[\operatorname{grad}\mu]) + s = \varrho\,\dot{\varphi}$$

- *Constitutive equations*

$$(\boldsymbol{\xi})_i = \varrho \partial_{\mathrm{grad}\,\varphi} \psi = \varrho \ell_{(2,1)} (\mathbf{y})_i$$

$$(\mathbf{S})_{ij} = 2\ell_{(4,1)} (\mathbf{B})_{ij}$$

—Summary of a second gradient theory—

Incompressible fluid flow with one conserved phase field:

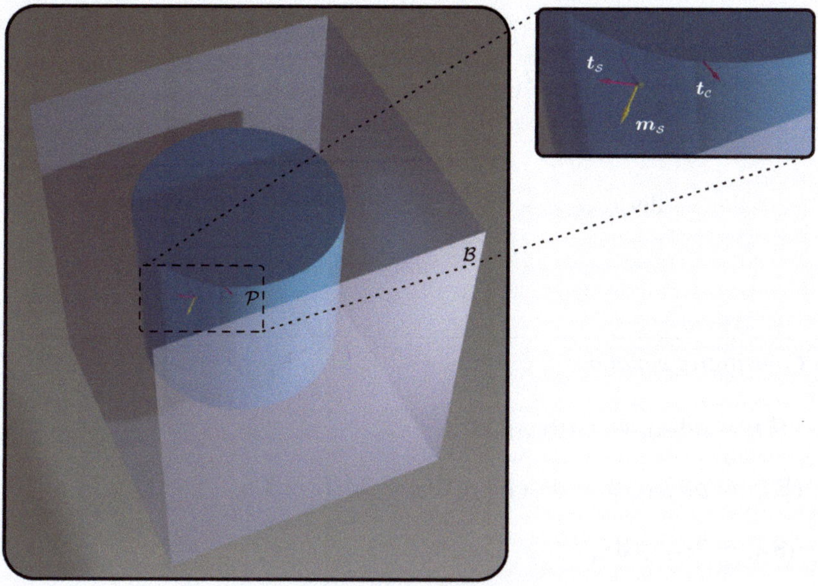

An arbitrary part of $\mathcal{B}$

• *Microtractions, tractions, couple-microtractions, and couple-tractions*

$$\xi_{\mathcal{S}} = (\boldsymbol{\xi} - \operatorname{div} \boldsymbol{\Xi}) \cdot \boldsymbol{n} - \operatorname{div}_{\mathcal{S}}(\mathbf{P}_{\!n}[\boldsymbol{\Xi}[\boldsymbol{n}]])$$

$$\boldsymbol{t}_{\mathcal{S}} = (\mathbf{T} - \operatorname{div} \mathbb{T})[\boldsymbol{n}] - \operatorname{div}_{\mathcal{S}}(\mathbb{T}[\boldsymbol{n}]\mathbf{P}_{\!n})$$

$$\boldsymbol{\varpi}_{\mathcal{S}} = (\boldsymbol{n} \otimes \boldsymbol{n})[\boldsymbol{\Xi}[\boldsymbol{n}]]$$

$$\boldsymbol{m}_{\mathcal{S}} = \boldsymbol{n} \times \mathbb{T}[\boldsymbol{n}][\boldsymbol{n}]$$

$$\xi_{\mathcal{C}} = \{\!\{\, \boldsymbol{\nu} \cdot \boldsymbol{\Xi}[\boldsymbol{n}] \,\}\!\}$$

$$\boldsymbol{t}_{\mathcal{C}} = \{\!\{\, \mathbb{T}[\boldsymbol{n}][\boldsymbol{\nu}] \,\}\!\}$$

• *Free-energy density*

$$\psi = f(\varphi) + \tfrac{1}{2}(\|\boldsymbol{y}\|^2 + \|\boldsymbol{\Upsilon}\|^2)$$

f: potential function, $\boldsymbol{y} = \operatorname{grad}\varphi$, $\boldsymbol{\Upsilon} = \operatorname{grad}^2\varphi$,
$\mathbf{B} := \operatorname{sym}(\operatorname{grad}\boldsymbol{v})$, and $\mathbb{B} := \operatorname{grad}^2\boldsymbol{v}$

• *Pointwise field equations*

$$\begin{aligned}\operatorname{div}(\mathbf{S} - \mathbf{1}g - (\operatorname{grad}\varphi \otimes \boldsymbol{\xi} + 2(\operatorname{grad}^2\varphi)\boldsymbol{\Xi}))\\ -\operatorname{div}^2(\mathbb{S} - \mathbf{1} \otimes \boldsymbol{g} - (\operatorname{grad}\varphi \otimes \boldsymbol{\Xi}))\\ +\boldsymbol{b}^{\mathrm{ni}} - \varrho\dot{\boldsymbol{v}} = \mathbf{0}\end{aligned}$$

$$\operatorname{div}\partial_{\operatorname{grad}\varphi}\psi - \operatorname{div}^2\partial_{\operatorname{grad}^2\varphi}\psi - \partial_\varphi\psi + \gamma/\varrho = -\mu$$

$$\operatorname{div}(\mathbf{M}[\operatorname{grad}\mu]) + s = \varrho\dot{\varphi}$$

• *Constitutive equations*

$$(\boldsymbol{\xi})_i = \varrho\partial_{\operatorname{grad}\varphi}\psi = \varrho\ell_{(2,1)}(\boldsymbol{y})_i$$

$$(\boldsymbol{\Xi})_{ij} = \varrho\partial_{\operatorname{grad}^2\varphi}\psi = \varrho\ell_{(4,1)}\delta_{ij}(\boldsymbol{\Upsilon})_{kk} + 2\varrho\ell_{(4,2)}(\boldsymbol{\Upsilon})_{ij}$$

$$(\mathbf{S})_{ij} = 2\ell_{(4,1)}(\mathbf{B})_{ij}$$

$$\begin{aligned}(\mathbb{S})_{ijk} = {}& 2\ell_{(6,1)}((\mathbb{B})_{kij} + (\mathbb{B})_{jik} - \tfrac{1}{2}(\delta_{ij}(\mathbb{B})_{kmm} + \delta_{ik}(\mathbb{B})_{jmm}))\\ &+ 2\ell_{(6,2)}(\mathbb{B})_{ijk} + \ell_{(6,3)}(\delta_{jk}(\mathbb{B})_{imm} - \tfrac{1}{2}(\delta_{ij}(\mathbb{B})_{kmm} + \delta_{ik}(\mathbb{B})_{jmm}))\end{aligned}$$

—Summary of a third gradient theory—

Incompressible fluid flow with one conserved phase field:

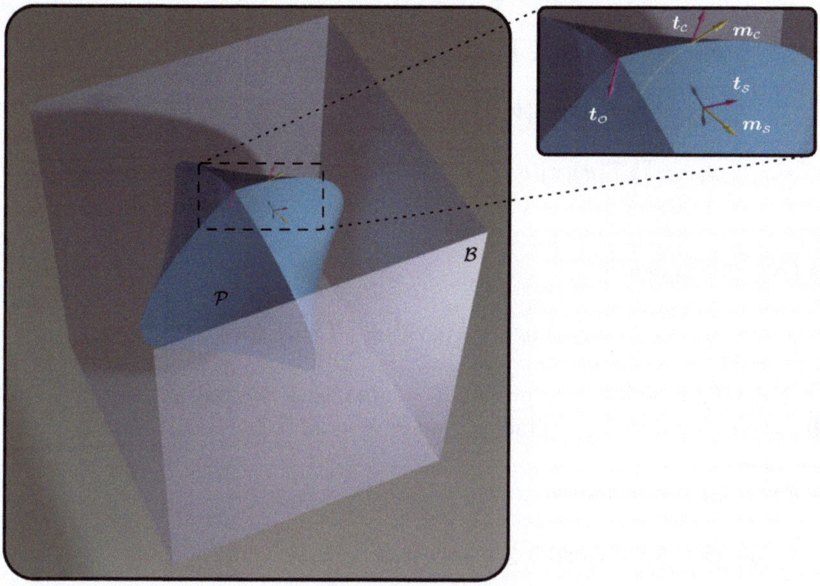

An arbitrary part of $\mathcal{B}$

• *Microforces, forces, microtractions, tractions, couple-microtractions, and couple-tractions*

$$\begin{aligned}\xi_{\mathcal{S}} &= (\boldsymbol{\xi} - \operatorname{div}\Xi + \operatorname{div}^2\mathbb{X})\cdot\boldsymbol{n} - \operatorname{div}_{\mathcal{S}}(\mathbf{P}_{\!n}[(\Xi - \operatorname{div}\mathbb{X})[\boldsymbol{n}]])\\ &\quad + \operatorname{div}_{\mathcal{S}}(\mathbf{P}_{\!n}[\operatorname{div}_{\mathcal{S}}(\mathbb{X}[\boldsymbol{n}]\mathbf{P}_{\!n})]) - \operatorname{div}_{\mathcal{S}}(\mathbb{X}[\boldsymbol{n}][\boldsymbol{n}]\cdot\mathbf{K}^{\top})\end{aligned}$$

$$\begin{aligned}\boldsymbol{t}_{\mathcal{S}} &= (\mathbf{T} - \operatorname{div}\mathbb{T} + \operatorname{div}^2\boldsymbol{\mathbb{T}})[\boldsymbol{n}] - \operatorname{div}_{\mathcal{S}}((\mathbb{T} - \operatorname{div}\boldsymbol{\mathbb{T}})[\boldsymbol{n}]\mathbf{P}_{\!n})\\ &\quad + \operatorname{div}_{\mathcal{S}}(\operatorname{div}_{\mathcal{S}}(\boldsymbol{\mathbb{T}}[\boldsymbol{n}]\mathbf{P}_{\!n})\mathbf{P}_{\!n}) - \operatorname{div}_{\mathcal{S}}(\boldsymbol{\mathbb{T}}[\boldsymbol{n}][\boldsymbol{n}]\mathbf{K}^{\top})\end{aligned}$$

$$\begin{aligned}\boldsymbol{\varpi}_{\mathcal{S}} &= (\boldsymbol{n}\otimes\boldsymbol{n})[(\Xi - \operatorname{div}\mathbb{X})[\boldsymbol{n}] - \operatorname{div}_{\mathcal{S}}(\mathbb{X}[\boldsymbol{n}]\mathbf{P}_{\!n})]\\ &\quad - \mathbb{X}[\boldsymbol{n}][\boldsymbol{n}]\cdot\mathbf{K}^{\top}\end{aligned}$$

$$\begin{aligned}\boldsymbol{m}_{\mathcal{S}} &= \boldsymbol{n}\times((\mathbb{T} - \operatorname{div}\boldsymbol{\mathbb{T}})[\boldsymbol{n}][\boldsymbol{n}] - \operatorname{div}_{\mathcal{S}}(\boldsymbol{\mathbb{T}}[\boldsymbol{n}]\mathbf{P}_{\!n})[\boldsymbol{n}])\\ &\quad - 2\operatorname{ax}(\operatorname{skw}(\boldsymbol{\mathbb{T}}[\boldsymbol{n}][\boldsymbol{n}]\mathbf{K}^{\top})^{\top})\end{aligned}$$

$$\begin{aligned}\xi_{\mathcal{C}} &= \{\!\!\{\, \boldsymbol{\nu}\cdot((\Xi - \operatorname{div}\mathbb{X})[\boldsymbol{n}] - \operatorname{div}_{\mathcal{S}}(\mathbb{X}[\boldsymbol{n}]\mathbf{P}_{\!n})\\ &\quad + \mathbb{X}[\boldsymbol{n}][\boldsymbol{n}]\cdot\mathbf{K}^{\top}) - \partial_{\sigma}(\boldsymbol{\sigma}\cdot\mathbb{X}[\boldsymbol{n}][\boldsymbol{\nu}])\}\!\!\}\end{aligned}$$

$$t_{\mathcal{C}} = \{\!\!\{\, (\mathbb{T} - \operatorname{div}\mathbf{\mathbb{T}})[\boldsymbol{n}][\boldsymbol{\nu}] - \operatorname{div}_{\mathcal{S}}(\mathbf{\mathbb{T}}[\boldsymbol{n}]\mathbf{P}_{\!n})[\boldsymbol{\nu}]$$

$$+ \mathbf{\mathbb{T}}[\boldsymbol{n}][\boldsymbol{n}][\mathbf{K}^{\top}[\boldsymbol{\nu}]] - \partial_{\sigma}(\mathbf{\mathbb{T}}[\boldsymbol{n}][\boldsymbol{\nu}][\boldsymbol{\sigma}])\}\!\!\}$$

$$\boldsymbol{\varpi}_{\mathcal{C}} = \{\!\!\{\, \mathbf{P}_{\!\sigma}[\mathbb{X}[\boldsymbol{n}][\boldsymbol{\nu}]]\, \}\!\!\}$$

$$\boldsymbol{m}_{\mathcal{C}} = 2\,\mathrm{ax}\,(\mathrm{skw}\,(\{\!\!\{\, \mathbf{\mathbb{T}}[\boldsymbol{n}][\boldsymbol{\nu}]\, \}\!\!\})^{\top}) - \boldsymbol{\sigma} \times \{\!\!\{\, \mathbf{\mathbb{T}}[\boldsymbol{n}][\boldsymbol{\nu}][\boldsymbol{\sigma}]\, \}\!\!\}$$

$$\xi_{\mathcal{O}} = [\![\{\!\!\{\, \boldsymbol{\sigma} \cdot \mathbb{X}[\boldsymbol{n}][\boldsymbol{\nu}]\, \}\!\!\}]\!]$$

$$\boldsymbol{t}_{\mathcal{O}} = [\![\{\!\!\{\, \mathbf{\mathbb{T}}[\boldsymbol{n}][\boldsymbol{\nu}][\boldsymbol{\sigma}]\, \}\!\!\}]\!]$$

- *Free-energy density*

$$\psi = f(\varphi) + \tfrac{1}{2}(\|\boldsymbol{y}\|^2 + \|\boldsymbol{\Upsilon}\|^2 + \|\mathbb{Y}\|^2)$$

f: potential function, $\boldsymbol{y} = \operatorname{grad}\varphi$, $\boldsymbol{\Upsilon} = \operatorname{grad}^2\varphi$, $\mathbb{Y} = \operatorname{grad}^3\varphi$,
$\mathbf{B} := \operatorname{sym}(\operatorname{grad}\boldsymbol{v})$, $\mathbb{B} := \operatorname{grad}^2\boldsymbol{v}$, and $\mathbf{\mathbb{B}} := \operatorname{grad}^3\boldsymbol{v}$

- *Pointwise field equations*

$$\begin{aligned}
&\operatorname{div}(\mathbf{S} - \mathbf{1}g - (\operatorname{grad}\varphi \otimes \boldsymbol{\xi} + 2(\operatorname{grad}^2\varphi)\boldsymbol{\Xi} + 3\operatorname{grad}^3\varphi : \mathbb{X}))\\
&\qquad -\operatorname{div}^2(\mathbb{S} - \mathbf{1} \otimes \boldsymbol{g} - (\operatorname{grad}\varphi \otimes \boldsymbol{\Xi} + 3\operatorname{grad}^2\varphi \cdot \mathbb{X}))\\
&\qquad\qquad +\operatorname{div}^3(\mathbf{\mathbb{S}} - \mathbf{1} \otimes \mathbf{G} - \operatorname{grad}\varphi \otimes \mathbb{X}))\\
&\qquad\qquad\qquad +\boldsymbol{b}^{\mathrm{ni}} - \varrho\dot{\boldsymbol{v}} = \mathbf{0}
\end{aligned}$$

$$\operatorname{div}\partial_{\operatorname{grad}\varphi}\psi - \operatorname{div}^2\partial_{\operatorname{grad}^2\varphi}\psi + \operatorname{div}^3\partial_{\operatorname{grad}^3\varphi}\psi - \partial_{\varphi}\psi + \gamma/\varrho = -\mu$$

$$\operatorname{div}(\mathbf{M}[\operatorname{grad}\mu]) + s = \varrho\dot{\varphi}$$

- *Constitutive equations*

$$(\boldsymbol{\xi})_i = \varrho\partial_{\operatorname{grad}\varphi}\psi = \varrho\ell_{(2,1)}(\boldsymbol{y})_i$$

$$(\boldsymbol{\Xi})_{ij} = \varrho\partial_{\operatorname{grad}^2\varphi}\psi = \varrho\ell_{(4,1)}\delta_{ij}(\boldsymbol{\Upsilon})_{kk} + 2\varrho\ell_{(4,2)}(\boldsymbol{\Upsilon})_{ij}$$

$$\begin{aligned}
(\mathbb{X})_{ijk} = \varrho\partial_{\operatorname{grad}^3\varphi}\psi &= 3\varrho\ell_{(6,1)}(\delta_{ij}(\mathbb{Y})_{kmm} + \delta_{ik}(\mathbb{Y})_{jmm} + \delta_{jk}(\mathbb{Y})_{imm})\\
&\quad + 6\varrho\ell_{(6,2)}(\mathbb{Y})_{ijk}
\end{aligned}$$

$$
\begin{aligned}
(\mathbf{S})_{ij} &= 2\ell_{(4,1)}(\mathbf{B})_{ij}\\
(\mathbb{S})_{ijk} &= 2\ell_{(6,1)}((\mathbb{B})_{kij} + (\mathbb{B})_{jik} - \tfrac{1}{2}(\delta_{ij}(\mathbb{B})_{kmm} + \delta_{ik}(\mathbb{B})_{jmm}))\\
&\quad + 2\ell_{(6,2)}(\mathbb{B})_{ijk} + \ell_{(6,3)}(\delta_{jk}(\mathbb{B})_{imm} - \tfrac{1}{2}(\delta_{ij}(\mathbb{B})_{kmm} + \delta_{ik}(\mathbb{B})_{jmm}))\\
(\mathbf{\mathbb{S}})_{ijkl} &= 3\ell_{(8,1)}((\delta_{ij}(\mathbf{\mathbb{B}})_{lknn} + \delta_{ij}(\mathbf{\mathbb{B}})_{klnn} + \delta_{ik}(\mathbf{\mathbb{B}})_{jlnn} + \delta_{ik}(\mathbf{\mathbb{B}})_{ljnn}\\
&\quad + \delta_{il}(\mathbf{\mathbb{B}})_{jknn} + \delta_{il}(\mathbf{\mathbb{B}})_{kjnn}) - 5((\mathbf{\mathbb{B}})_{lijk} + (\mathbf{\mathbb{B}})_{jikl} + (\mathbf{\mathbb{B}})_{kijl}))\\
&\quad + 3\ell_{(8,2)}(\delta_{jk}(\mathbf{\mathbb{B}})_{limm} + \delta_{jl}(\mathbf{\mathbb{B}})_{kimm} + \delta_{kl}(\mathbf{\mathbb{B}})_{jimm}\\
&\quad - ((\mathbf{\mathbb{B}})_{lijk} + (\mathbf{\mathbb{B}})_{jikl} + (\mathbf{\mathbb{B}})_{kijl}))\\
&\quad + 3\ell_{(8,3)}(\delta_{kl}(\mathbf{\mathbb{B}})_{ijnn} + \delta_{jl}(\mathbf{\mathbb{B}})_{ikmm} + \delta_{jk}(\mathbf{\mathbb{B}})_{ilmm}\\
&\quad - ((\mathbf{\mathbb{B}})_{lijk} + (\mathbf{\mathbb{B}})_{jikl} + (\mathbf{\mathbb{B}})_{kijl}))\\
&\quad + 6\ell_{(8,4)}(\mathbf{\mathbb{B}})_{ijkl}
\end{aligned}
$$

1 Final Remarks

In this study, I aimed to describe the mechanics of enriched continua. Focusing on gradient theories, I accounted for motion accompanied by various transition layers. I based my treatment on the principle of virtual powers to determine the explicit form of the fields present in the external virtual power balance and obtain the pointwise version of the balances of microforces and forces. In the internal virtual power, I incorporated the first, second, and third gradients of the phase fields and motion. These gradients arise by considering various kinematic processes which are intimately connected to the level of the arbitrariness of the Euler–Cauchy cuts. The surface defining the Euler–Cauchy cut may lose its smoothness along a curve, namely, junction-edge. A junction-edge may also lose its smoothness at a point, namely, junction-point. Guided by the presence of gradients up to third order in conjunction with this level of arbitrariness in the virtual power principle, I was led to define on each geometrical feature the surface microtractions/tractions, the surface hypermicrotractions/hypertractions, the surface supra-hypermicrotractions/hypertractions, the junction-edge microtractions/tractions, the junction-edge hypermicrotractions/hypertractions, and the junction-points microforces/forces. These quantities are the result of different types of interactions between adjacent parts of the body. The pointwise balances are then integrated on an arbitrary part to arrive at the partwise balances of microforces, forces, microtorques, and torques. Next, I postulated surface balances of microforces, forces, microtorques, and torques. Additionally, I postulated the principle of virtual power on surfaces. Then, the first and second laws of thermodynamics with the power balance provided me with suitable and consistent choices for the constitutive equations. Finally, the complementary balances, namely, the balances on surfaces, were taken to the limit of the body to coincide with different parts of the boundaries of the body. These surface balances were then called environmental surface balances and aided in determining suitable and consistent boundary conditions. Ultimately, the environmental surface power balance is relaxed to yield an environmental surface imbalance of powers, rendering a more general type of boundary condition.

There are uncountable relevant considerations dismissed in this study. Here, I list a few of them.

- Consider inertial components into surface, junction-edge, and junction-point microtractions/tractions;
- Prove the existence of supra hypermicrostresses/hyperstress fields in a variational manner, à la Fosdick [23, 24];
- Thorough discussion on the equivalence between the power expended by hypertractions and couples fields;
- Study the implications arising in the representation theorem for eighth-order isotropic tensors when working in three-dimensions instead of four;
- Discussion on strong and weak no-slip conditions;
- Analytical and manufactured solutions in simple problem configurations;
- Numerical implementation.

Bibliography

1. Espath, L.: A continuum framework for phase field with bulk-surface dynamics. Partial Differ. Equ. Appl. **4**(1), 1–17 (2023)
2. Gurtin, M.E., Fried, E., Anand, L.: The Mechanics and Thermodynamics of Continua. Cambridge University Press (2010)
3. Malvern, L.E.: Introduction to Continuum Mechanics. Prentice Hall Inc., Engle Cliffs, NJ, 1696(5):0 (1969)
4. Morton, E.G.: An Introduction to Continuum Mechanics. Academic Press (1982)
5. Kemp, M.: Leonardo da vinci's laboratory: studies in flow. Nature **571**(7765), 322–324 (2019)
6. Marusic, I., Broomhall, S.: Leonardo da vinci and fluid mechanics. Ann. Rev. Fluid Mech. **53**, 1–25 (2021)
7. Cosserat, E., Cosserat, F.: Théorie des corps déformables. A. Hermann et fils (1909)
8. Maugin, G.A., Metrikine, A.V.: Mechanics of Generalized Continua. Springer (2010)
9. Voight, W.: Theoretical Studies on the Elasticity Ratios of Crystals, I. II. Dependence. Ges. Wiss, Gottingen (1887)
10. Ganghoffer, J.-F.: Cosserat, Eugène and François. In: Altenbach, H., Öchsner, A. (eds.) Encyclopedia of Continuum Mechanics, pp. 1–6. Springer, Berlin (2017)
11. Edelen, D.G.B.: A global formulation of continuum physics and the resulting equivalence classes of nonlocal field equations. In: Rogula, D. (ed) Nonlocal Theories of Material Systems, pp. 41–79. Ossolineum, Wrocław, Poland (1976)
12. Duhem, P.: Le potentiel thermodynamique et la pression hydrostatique. In: Annales scientifiques de l'École Normale Supérieure **10**, 183–230 (1893)
13. Ericksen, J.L., Truesdell, C.: Exact theory of stress and strain in rods and shells. Arch. Ration. Mech. Anal. **1**(1), 295–323 (1957)
14. Ericksen, J.L.: Liquid crystals with variable degree of orientation. Arch. Ration. Mech. Anal. **113**(2), 97–120 (1991)
15. Toupin, R.A.: Elastic materials with couple-stresses. Arch. Ration. Mech. Anal. **11**(1), 385–414 (1962)
16. Toupin, R.A.: Theories of elasticity with couple-stress. Arch. Ration. Mech. Anal. **17**(2), 85–112 (1964)
17. Mindlin, R.D., Eshel, N.N.: On first strain-gradient theories in linear elasticity. Int. J. Solids Struct. **4**(1), 109–124 (1968)
18. Germain, P.: La méthode des puissances virtuelles en mécanique des milieux continus. J. Mécanique **12**, 236–274 (1973)

L. Espath, *Mechanics and Geometry of Enriched Continua*,
https://doi.org/10.1007/978-3-031-28934-7

19. Germain, P.: The method of virtual power in continuum mechanics. part 2: Microstructure. SIAM J. Appl. Math. **25**(3), 556–575 (1973)
20. Fried, E., Gurtin, M.E.: Tractions, balances, and boundary conditions for nonsimple materials with application to liquid flow at small-length scales. Arch. Ration. Mech. Anal. **182**(3), 513–554 (2006)
21. Noll, W., Virga, E.G.: On edge interactions and surface tension. Arch. Ration. Mech. Anal. **111**(1), 1–31 (1990)
22. Dell'Isola, F., Seppecher, P.: Edge contact forces and quasi-balanced power. Meccanica **32**(1), 33–52 (1997)
23. Fosdick, R.L., Virga, E.G.: A variational proof of the stress theorem of Cauchy. Arch. Ration. Mech. Anal. **105**(2), 95–103 (1989)
24. Fosdick, R.: A generalized continuum theory with internal corner and surface contact interactions. Continuum Mech. Thermodyn. **28**(1–2), 275 (2016)
25. Van der Waals, J.D.: The thermodynamic theory of capillarity under the hypothesis of a continuous variation of density. J. Stat. Phys. **20**(2), 200–244 (1979)
26. Korteweg, D.J.: Sur la forme que prennent les équations du mouvements des fluides si l'on tient compte des forces capillaires causées par des variations de densité considérables mais connues et sur la théorie de la capillarité dans l'hypothèse d'une variation continue de la densité. Archives Néerlandaises des Sciences exactes et naturelles **6**, 1–24 (1901)
27. Dunn, J.E., Serrin, J.: On the thermomechanics of interstitial working. Arch. Ration. Mech. Anal. **88**(2), 95–133 (1985)
28. Anderson, D.M., McFadden, G.B., Wheeler, A.A.: Diffuse-interface methods in fluid mechanics. Ann. Rev. Fluid Mech. **30**(1), 139–165 (1998)
29. Fried, E., Gurtin, M.E.: Continuum theory of thermally induced phase transitions based on an order parameter. Phys. D: Nonlinear Phenomena **68**(3–4), 326–343 (1993)
30. Fried, E., Gurtin, M.E.: Dynamic solid-solid transitions with phase characterized by an order parameter. Phys. D: Nonlinear Phenomena **72**(4), 287–308 (1994)
31. Gurtin, M.E.: Generalized Ginzburg-Landau and Cahn-Hilliard equations based on a microforce balance. Phys. D: Nonlinear Phenomena **92**(3–4), 178–192 (1996)
32. Espath, L., Calo, V.: Phase-field gradient theory. Zeitschrift fuer Angewandte Mathematik und Physik (ZAMP) **72**(45), 1–33 (2021)
33. Espath, L., Calo, V.M., Fried, E.: Generalized Swift-Hohenberg and phase-field-crystal equations based on a second-gradient phase-field theory. Meccanica **55**(10), 1853–1868 (2020)
34. Gurtin, M.E., Polignone, D., Vinals, J.: Two-phase binary fluids and immiscible fluids described by an order parameter. Math. Models Methods Appl. Sci. **6**(06), 815–831 (1996)
35. Green, A.E., Rivlin, R.S.: Simple force and stress multipoles. Arch. Ration. Mech. Anal. **16**, 325–353 (1964)
36. Green, A.E., Rivlin, R.S.: Multipolar continuum mechanics. Arch. Ration. Mech. Anal. **17**, 113–147 (1964)
37. Svendsen, B., Bertram, A.: On frame-indifference and form-invariance in constitutive theory. Acta Mech. **132**(1), 195–207 (1999)
38. Bertram, A., Svendsen, B.: On material objectivity and reduced constitutive equations. Arch. Mech. **53**(6), 653–675 (2001)
39. Svendsen, B.: On thermodynamic-and variational-based formulations of models for inelastic continua with internal lengthscales. Comput. Methods Appl. Mech. Eng. **193**(48–51), 5429–5452 (2004)
40. Svendsen, B.: Continuum thermodynamic and rate variational formulation of models for extended continua. In: Advances in Extended and Multifield Theories for Continua, pp. 1–18. Springer (2011)
41. Svendsen, B., Neff, P., Menzel, A.: On constitutive and configurational aspects of models for gradient continua with microstructure. ZAMM-J. Appl. Math. Mech./Zeitschrift für Angewandte Mathematik und Mechanik: Appl. Math. Mech. **89**(8), 687–697 (2009)
42. Gurtin, M.E.: A gradient theory of single-crystal viscoplasticity that accounts for geometrically necessary dislocations. J. Mech. Phys. Solids **50**(1), 5–32 (2002)

43. Podio-Guidugli, P.: Contact interactions, stress and material symmetry. Theor. Appl. Mech. **28–29**(28–29), 261–276 (2002)
44. Carillo, S., Podio-Guidugli, P., Caffarelli, G.V.: Second-order surface potentials in finite elasticity. In: Rational Continua, Classical and New, pp. 19–38. Springer (2003)
45. Espath, L.: On the control volume arbitraryness in the Navier-Stokes equation. Phys. Fluids **33**(1), 015110 (2021)
46. Clavijo, S.P., Espath, L., Calo, V.: Extended Larché-Cahn framework for reactive Cahn-Hilliard multicomponent systems. Continuum Mech. Thermodyn. **33**, 2391–2410 (2021)
47. Clavijo, S.P., Espath, L., Calo, V.: A continuum theory for mineral solid solutions undergoing chemo-mechanical processes. Continuum Mech, Thermodyn (2021)
48. Podio-Guidugli, P., Vianello, M.: Hypertractions and hyperstresses convey the same mechanical information. Continuum Mech. Thermodyn. **22**(3), 163–176 (2010)
49. Biria, A., Maleki, M., Fried, E.: Continuum theory for the edge of an open lipid bilayer. In: Advances in Applied Mechanics, vol. 46, pp. 1–68. Elsevier (2013)
50. Capecchi, D.: History of Virtual Work Laws: A History of Mechanics Prospective, vol. 42. Springer Science & Business Media (2012)
51. Lidström, P.: On the principle of virtual power in continuum mechanics. Math. Mech. Solids **17**(5), 516–540 (2012)
52. Del Piero, G.: On the method of virtual power in continuum mechanics. J. Mech. Mater. Struct. **4**(2), 281–292 (2009)
53. Truesdell, C., Noll, W.: The non-linear field theories of mechanics. In: The Non-Linear Field Theories of Mechanics, pp. 1–579. Springer (2004)
54. Larche, F.C., Cahn, J.W.: Thermochemical equilibrium of multiphase solids under stress. Acta Metallurgica **26**(10), 1579–1589 (1978)
55. Eliot, F., Morton, E.G.: Coherent solid-state phase transitions with atomic diffusion: A thermomechanical treatment. J. Stat. Phys. **95**(5), 1361–1427 (1999)
56. Dresselhaus, M.S., Dresselhaus, G., Jorio, A.: Group Theory: Application to the Physics of Condensed Matter. Springer Science & Business Media (2007)
57. Kearsley, E.A., Fong, J.T.: Linearly independent sets of isotropic cartesian tensors of ranks up to eight. J. Res. Natl. Bur. Stan. 49–58 (1975)
58. Gurtin, M.E., Vargas, A.S.: On the classical theory of reacting fluid mixtures. Arch. Ration. Mech. Anal. **43**, 179–197 (1997)
59. Coleman, B.D., Noll, W.: The thermodynamics of elastic materials with heat conduction and viscosity. Arch. Ration. Mech. Anal. **13**(1), 167–178 (1963)
60. Fried, E., Gurtin, M.E.: Thermomechanics of the interface between a body and its environment. Continuum Mech. Thermodyn. **19**(5), 253–271 (2007)
61. Duda, F.P., Sarmiento, A., Fried, E.: Phase fields, constraints, and the Cahn-Hilliard equation. Meccanica **56**, 1707–1725 (2021)
62. Fischer, H.P., Maass, P., Dieterich, W.: Novel surface modes in spinodal decomposition. Phys. Rev. Lett. **79**(5), 893 (1997)
63. Goldstein, G.R., Miranville, A., Schimperna, G.: A Cahn-Hilliard model in a domain with non-permeable walls. Phys. D: Nonlinear Phenomena **240**(8), 754–766 (2011)
64. Knopf, P., Lam, K.F., Liu, C., Metzger, S.: Phase-field dynamics with transfer of materials: The Cahn–Hilliard equation with reaction rate dependent dynamic boundary conditions. ESAIM: Math. Model. Numer. Anal. **55**(1), 229–282 (2021)

MIX
Papier aus verantwortungsvollen Quellen
Paper from responsible sources
FSC® C105338

If you have any concerns about our products,
you can contact us on
ProductSafety@springernature.com

In case Publisher is established outside the EU,
the EU authorized representative is:
Springer Nature Customer Service Center GmbH
Europaplatz 3, 69115 Heidelberg, Germany

Printed by Libri Plureos GmbH
in Hamburg, Germany